CAMBRIDGE SERIES ON HUMAN-COMPUTER INTERACTION 1

Cognitive Ergonomics and
Human-Computer Interaction

Cambridge Series on Human-Computer Interaction

Managing Editor: Professor J. Long,
Ergonomics Unit, University College, London.

Titles in the Series
1. J. Long and A. Whitefield *Cognitive Ergonomics and Human-Computer Interaction*
2. M. Harrison and H. Thimbleby *Formal Methods in Human-Computer Interaction*

Cognitive Ergonomics and Human-Computer Interaction

Edited by J. LONG and A. WHITEFIELD

Ergonomics Unit, University College, London

The right of the
University of Cambridge
to print and sell
all manner of books
was granted by
Henry VIII in 1534.
The University has printed
and published continuously
since 1584.

CAMBRIDGE UNIVERSITY PRESS

Cambridge

New York   Port Chester

Melbourne   Sydney

CAMBRIDGE UNIVERSITY PRESS
Cambridge, New York, Melbourne, Madrid, Cape Town,
Singapore, São Paulo, Delhi, Tokyo, Mexico City

Cambridge University Press
The Edinburgh Building, Cambridge CB2 8RU, UK

Published in the United States of America by Cambridge University Press, New York

www.cambridge.org
Information on this title: www.cambridge.org/9780521204842

First published 1989
First paperback edition 2011

A catalogue record for this publication is available from the British Library

ISBN 978-0-521-37179-7 Hardback
ISBN 978-0-521-20484-2 Paperback

Contents

Contents

Contributors

Phil Barnard; MRC Applied Psychology Unit, 15 Chaucer Road, Cambridge CB2 2EF, UK.

Richard Bornat; Computer Systems Laboratory, Queen Mary College, London E1 4NS, UK.

Paul Buckley; Department of Computer Science, Queen Mary College, University of London, Mile End Road, London E1 4NS, UK.

John Campion; AXC division, Admiralty Research Establisment, Portsdown, Portsmouth PO6 4AA, UK.

Dan Diaper; Department of Computer Science, Chadwick Building, The University of Liverpool, PO Box 147, Liverpool L69 3BX, UK.

Jonathan Grudin; Microelectronics and Computer Technology Corporation, Austin, Texas, USA.

Peter Johnson; Department of Computer Science, Queen Mary College, University of London, Mile End Road, London E1 4NS, UK.

John Long; Ergonomics Unit, University College London, 26 Bedford Way, London WC1H 0AP, UK.

Allan Maclean; MRC Applied Psychology Unit, 15 Chaucer Road, Cambridge CB2 2EF, UK.

Stephen J. Payne; Departments of Psychology and Computing, University of Lancaster, LA1 4YF, UK.

Harold Thimbleby; Department of Computing Science, University of Stirling, Stirling, FK9 4LA, UK.

Andy Whitefield; Ergonomics Unit, University College London, 26 Bedford Way, London WC1H 0AP, UK.

Preface

Computer technology and its applications are growing rapidly. Computers continue to extend both the range of human activities they affect and the effects they produce. These effects concern the users of computers and the domains of work to which the computer is applied. But this development has not been without its problems. One set of problems concerns the relationship between the user and the computer in interactive systems. Issues of user well-being and of system effectiveness and productivity have been major concerns underlying the developing interest in human-computer interaction. The problems originate with different people: the end-users who directly experience the consequences of poorly designed systems; the developers who make decisions about system interfaces; the engineers concerned with formulating principles, methods and tools for system developers; and the scientists interested in explaining and predicting the behaviour of human-computer systems. Human-computer interaction is fast becoming an established subject recruiting multi-disciplinary skills and knowledge to real-world applications.

Associated with this growth in human-computer interaction has been the growth of cognitive ergonomics – the application of the cognitive sciences to problems of human-machine interaction. Although the conceptual and empirical tools of cognitive ergonomics are applicable to many non-computer systems, human-computer interaction has provided a subject and a set of problems which have been a crucial stimulus to the development of cognitive ergonomics.

The aim of this book is to present some examples of cognitive ergonomics and their contributions to the development of human-computer interaction. Moreover, we have tried to identify and to contextualise the different kinds of activities that are presented, primarily in terms of the kinds of support cognitive ergonomics can offer human-computer interaction. By so doing, we hope to contribute to the development of both cognitive ergonomics and human-computer interaction.

The selection of topics for the chapters was made, in conjunction with the authors, in an attempt to meet certain goals. We sought reports and discussion of research (both scientific and engineering) and particularly of substantive, completed research. Each chapter (except the first) is based on theoretical and empirical research activity, in some cases extending for several years. Their conclusions and proposals therefore have solid foundations. The first chapter, in contrast, provides a framework for human-computer interaction and cognitive ergonomics. Its function is to provide coherence for the contributions and to facilitate the identification of similarities and differences between the approaches proposed by the contributions. In addition, part of the framework is used to locate all the contributions relative to one another.

A further criterion for selection was to include examples of the full range of activities in cognitive ergonomics, in order to exemplify the various types of support for human-computer interaction. Although we have been generally successful, there is probably a greater representation of knowledge acquisition activities than application activities. The use of the framework in Chapter 1 is a major part of our attempt to produce a book whose various contents are integrated into a coherent whole. This goal of coherence, along with the wish to produce a book of high quality, required strong editing, and accordingly, each chapter has been thoroughly considered by both editors through two or more versions. The final say as to its contents, however, has always been the author's.

We anticipate that this book will be most suited for a research readership, where research is interpreted broadly to include science and engineering activities, both of research and system development. Given the applied nature and industrial relevance of human-computer interaction, this readership is likely to go well beyond academic and research institutions. The book is likely to have little direct benefit to students in the early stages of first degree courses, to end-users of interactive systems, and to those system developers who are unconcerned with theories, methods and principles underlying the improvement of interactive systems. It will, however, be of interest to any of these groups who are concerned about the problems of human-computer interaction.

Our thanks are due to the various people involved in producing this book. First and foremost are the authors, who put a great deal of effort into their chapters and who responded positively, and often gracefully, to our considerable editorial demands. In addition, we wish to thank: Dan Diaper and Peter Johnson, who were involved in early discussions about the contents of the book; Andrew Life and John Dowell, who provided additional editing of some chapters; David Tranah, Susan Milmoe and other staff at CUP, particularly for their efforts in launching the series along with this book; the personnel at the Ergonomics Unit, University College London, for their forbearance during the book's production; and Doris Long for typing many of the editorial comments.

To conclude, we believe that the issues surrounding human-computer interaction are intellectually very stimulating and practically very important, and that cognitive ergonomics provides an appropriate position from which to approach them. We hope this book demonstrates some substance to these beliefs.

John Long and Andy Whitefield
London, July 1988

1

Cognitive Ergonomics and Human-Computer Interaction

An introduction

John Long

1.1 Introduction

The aim of this book is to present some examples of cognitive ergonomic contributions to human-computer interaction and to associate them within a common and coherent perspective. Human-computer interaction seeks to support interactions between humans and computers which make computers effective for performing work. Cognitive ergonomics contributes to human-computer interaction by supporting those aspects of the interactions which depend on the knowledge, required by the human, to use the computer to perform work effectively. The aim of this chapter is to introduce the examples of cognitive ergonomics and to provide the common perspective in the form of a framework. A second aim is to describe how part of the framework serves to organise the examples within the book.

The chapter presents in Sections 1.2 and 1.3 a general characterisation of the phenomena and discipline of human-computer interaction and the contribution which cognitive ergonomics makes to them. Section 1.4 reviews the examples of cognitive ergonomics and identifies their requirements for a framework. These are: concepts; methods; interaction development practice; and support for interaction development practice. Section 1.5 then proposes a framework which addresses these requirements, first in terms of human-computer interaction, and second in terms of the contribution of cognitive ergonomics to it. The framework is used to identify the similarities and differences between the examples. Finally, Section 1.6 describes how part of the framework – the science support for interaction development practice – is used to organise the examples, and suggests how they may be related in terms of this support.

Reading this chapter first, then, should facilitate an understanding of the examples of cognitive ergonomics which follow. However, consideration should be given to reading it a second time – after the other chapters – to gain a better understanding of the discussion of the examples presented in Section 1.5. In addition, readers experiencing difficulty with Sections 1.4 and 1.5 – perhaps because they are new to the area – should consider omitting them on first reading and proceeding from Section 1.3 to Section 1.6.

1.2 Human-computer interaction

Human-computer interaction comprises phenomena and a discipline which takes those phenomena as its scope. The phenomena involve systems consisting of: people – both as individuals and as social organisations; computers – both stand-alone and as networks; and their interaction. Since the systems are physical and informational, so too are their interactions. The discipline is concerned to support the optimisation of the interaction between humans and computers to perform work effectively. The concern, then, is not with the interactions in isolation. Humans use computers to do work and also have performance requirements for the work which is carried out. Interactions and their optimisation, then, need to be developed in the context of work and performance. It follows that human-computer interaction as a discipline aims to support the optimisation of human-computer interactions for effectiveness (Dowell and Long, 1988b).

The discipline of human-computer interaction consists of two main sub-disciplines. The first, termed software engineering (and computer science) primarily concerns the computer side of the interaction (that is, in the context of human-computer interaction). Software engineering aims to develop specifications (for both hardware and software aspects of the computer) which can be implemented as an interaction. The second, termed ergonomics (or human factors), primarily concerns the human side of the interaction. Ergonomics aims to develop specifications (for both physical and mental aspects of the human) which can be implemented as an interaction. Both sub-disciplines address the interaction, although with different but complementary emphases. They both aim to support the optimisation of the interaction for effectiveness. Cognitive ergonomics is concerned with the mental aspects of the interaction and so with developing specifications of the knowledge required by the human to interact with the computer to perform work effectively. These specifications are implementable as an interaction.

1.3 Cognitive ergonomics

To support the specification of knowledge required by humans to use computers, the discipline of cognitive ergonomics itself needs to acquire and apply principles concerning that knowledge. Knowledge involved in computer-based work can be conceived as consisting of representations and processes: representations of the work to be performed, of the computer, etc; and processes required to use the representations to perform the work, to use the computer, etc. The processes include the acquisition of the representations, their transformation and their expression in the form of behaviour. Cognitive ergonomics, then, seeks to support the specification of knowledge conceived as the representations and processes required by the humans to interact with the computer to perform work effectively. Cognitive compatibility between the representations and the processes required by the interaction and those currently possessed or acquirable by the human has been advanced as one principle on which specifications can be based (Long, 1987). The principle typifies the support offered by cognitive ergonomics as a discipline.

The disciplines of cognitive science contribute generally to cognitive ergonomics. Some disciplines, such as mathematics, logic and computing, provide formal systems for the expression of representations and processes. Other disciplines, such as cognitive psychology, artificial intelligence and linguistics, seek to explain and predict behaviour in terms of formal or semi-formal systems. Together, the analytic and empirical support, provided by these scientific disciplines for representations and processes, contribute generally to the capability of cognitive ergonomics to develop specifications of interactions whose implementations achieve some desired effectiveness of work.

Cognitive ergonomic support for the development of specifications of the knowledge required by interactions may take different forms. Selection tests for computer users, for example, aim to specify the knowledge required by interactive behaviour. Aptitude tests aim to specify the potential for user-knowledge to be extended. Training programmes and 'help' facilities specify the additions to user-knowledge required by the interaction. Lastly, specifying the interface, including the software associated with the interaction, also dictates the knowledge required to use the computer. Although cognitive ergonomic contributions to the discipline of human-computer interaction may take many forms, only some of which are formally expressed at present, they can all be understood as supporting the development of implementable specifications of knowledge required by the human to interact with the computer to perform work effectively. The chapters of this book all offer contributions to human-computer interaction, although in a variety of forms.

1.4   Requirements for a framework

The function of the framework which follows is to support a common and coherent perspective for the cognitive ergonomic examples of human-computer interaction presented in this book. The framework provides coherence by offering an analytic structure which can be used to model examples of different kinds. The common form of description, which it affords, aids identification of the similarities and differences between them, and so facilitates comparison. Comparisons are instructive, because they make possible evaluation of the appropriateness of examples which offer alternative approaches. Further, the framework helps to ensure the completeness of the comparison.

Any discipline must at least have as its scope the concepts and methods of the practice it supports. The framework for the contributions, then, should include the following: cognitive ergonomic concepts and methods; cognitive ergonomic interaction development practice and support for interaction development practice. Specifying the knowledge required by humans to interact with computers is a practice involving many methods. Practice may gain from support of different kinds. Concepts and methods are required by both interaction development practice and associated support activities. Concepts, methods, interaction development practice, and support for practice, then, need to be included in the framework. Each of the requirements will be exemplified in turn from the examples presented in this book.

First, the examples draw on a broad range of *concepts* which describe and relate humans, computers, interaction and knowledge. These include concepts associated with the entities themselves, for example, 'device', 'system', 'program' for the computer (see Payne – Chapter 5; Bornat and Thimbleby – Chapter 8). Likewise with the actions of the entities, for example, 'activities', 'behaviour', 'performance' for the human (see Barnard et al. – Chapter 4; Buckley – Chapter 6). Other concepts are associated with the interaction, for example, 'task', 'control' and 'interface' (see Campion – Chapter 2; Whitefield – Chapter 3). Likewise with the knowledge used to support the interaction, for example, 'representations', 'models', and 'methods' (see Diaper and Johnson – Chapter 7). Some concepts are more coherently and completely defined than others – for example, 'device' more than 'system'. Some concepts are more similarly used than others – for example, 'activity' more than 'performance'. However, many concepts remain undefined and their usage varies with the contribution, for example 'behaviour' and 'task'. The first requirement for the framework, then, is to suggest how these concepts might be coherently related.

The second requirement for the framework is to accommodate the different *methods* used by cognitive ergonomics. Some methods are described explicitly, for example, how to create syllabi for training students to interact with computers (see Diaper and Johnson – Chapter 7). Other methods are more implicit, for example, how to design a text display editor (see Bornat and Thimbleby – Chapter 8). In addition, some methods provide notations to aid reasoning about the knowledge required by a human to effect an interaction (see Payne – Chapter 5). Other methods involve the collection of data by observational or experimental means to aid such reasoning (see Barnard et al. – Chapter 4). It must be possible to use the framework to model the different methods exemplified by the contributions.

The third requirement is to accommodate the different forms of *interaction development practice* described or implied by the contributions. Some forms of development are informal (see Buckley – Chapter 6). Others involve the implementation of an interface to support the interaction early in the development process (see Bornat and Thimbleby – Chapter 8). In addition, the practice of cognitive ergonomics in the development may vary relative to that of software engineering. Specifying the knowledge required by the interaction may be effected entirely by software engineers/computer scientists (see Bornat and Thimbleby – Chapter 8). Alternatively, the practice of cognitive ergonomic interaction development may evaluate the operational interface and suggest how it might be redesigned to improve the effectiveness of the interaction (see Buckley – Chapter 6). The forms of intended practice aids may also vary – from an analytic design tool (see Whitefield – Chapter 3) to a design procedure (see Campion – Chapter 2). The framework must be able to model these different aspects of the practice of cognitive ergonomic interaction development.

The last requirement for the framework is to accommodate *support for interaction development practice*. Some support involves the construction of a science base (see Barnard et al. – Chapter 4). Different support includes the development an interface (see Bornat and Thimbleby – Chapter 8). Yet other contributions describe support intended to develop principles to be embodied in future interfaces (see Campion – Chapter 2). The framework, then, needs to be able to model these different forms of support for the practice of cognitive ergonomic interaction development. The framework is presented in the following section and addresses each of these requirements in turn.

1.5   A framework for the examples

The framework addresses the four requirements: concepts (in Section 1.5.1); methods (in Section 1.5.2); interaction development practice (in Section 1.5.3); and support for interaction development practice (in Section 1.5.4). The requirements are addressed, first in terms of human-computer interaction, and then in terms of the contribution of cognitive ergonomics to it. The framework is applied by means of examples drawn from the contributions: examples of concepts (in Section 1.5.1); examples of methods (in Section 1.5.2); examples of interaction development practice (in Section 1.5.3); and examples of support for interaction development practice (in Section 1.5.4).

1.5.1   Concepts

The framework for the concepts of human-computer interaction originates from an engineering conception proposed by Dowell and Long (1988a,b) for the purpose of specifying the problem of evaluating user-computer interactions at an early stage of a system's development. The conception distinguishes: work; the entities which perform work; and the effectiveness with which the entities perform work.

In the conception, *work* occurs within and is distributed by *organisations*. Work is performed in a real world of objects which are the scope of some intent or requirement. Objects are characterised by their attributes and relations where those attributes afford a potential for change. That potential is realised by the *task*. The potential for change of attribute states of a class of objects is termed an *application domain* and is principally determined by organisational and technological constraints. Organisations allocate tasks in terms of *task goals* which are expressed as desired, that is intended, states of objects. Intentional attainment of the desired state constitutes an ideal outcome of the task with respect to its goal. However, since many different compromise outcomes to tasks are possible, *quality* describes how they can be compared and assessed. In summary, tasks are changes required, or intentionally produced, in the objects constituting the intended world in which work occurs.

Organisations assign work to systems. A work *system* is a set of mutually influential entities associated for the purpose of producing desired changes in the attribute state of objects. The system performs tasks, achieving task goals within application domains. Human-computer interaction is concerned with systems whose entities are *humans* and *computers*. The computer is considered to be an *Instrument* deployed by a *user* to perform tasks. *Online* tasks produce changes in objects represented by the system. *Offline* tasks produce changes in objects not so represented. *Interaction* is the mutual influence of *user behaviour* (or part thereof) and of *computer behaviour* (or part thereof) and so determines *system behaviour*. Human-computer interaction is concerned with the optimisation of user and computer behaviours and includes the user-computer *interface*. This interface is constituted by all the structural variables of both the user and the computer determining the behaviour of the system in its online tasks.

Performance expresses the effectiveness of a system relating the quality of the task product and the cost incurred in its production. Performance is different from, but determined by, behaviour. Behaviour may be more or less optimal with respect to performance. The general problem of human-computer interaction, earlier expressed informally as 'optimising human-computer interactions for the effective performance of work', can now be expressed in terms of the conception as: 'Specification and implementation of computer behaviour and user behaviour such that user behaviour, interacting with computer behaviour, produces desired performance.' (A more complete description can be found in Dowell and Long, 1988a,b.)

The contribution of ergonomics or human factors to the solution of the general problem of human-computer interaction is to develop implementable specifications of user behaviour with respect to the interface, broadly defined such that interactions produce desired performance. The contribution of cognitive ergonomics is to develop the specification of the knowledge which determines user behaviour. Knowledge in cognitive ergonomics is typically expressed in terms of the mental representations and processes employed by the user to interact with the computer. Representations are of the applications domain, the task, the computer, etc. Processes would include those involved in the acquisition of representations, their transformation, and their expression in behaviour (see Section 1.3). Taken together, the concepts of cognitive ergonomics and those of human-computer interaction constitute the framework which can be applied to the examples presented here.

Examples of the concepts No contribution provides a complete illustration of all the concepts described in the framework. That is to say, none of the examples expresses or supports the specification of the mental representations and processes required to produce that user behaviour which, when interacting via the interface with computer behaviour, engenders desired system performance. None expresses performance in terms of the quality of change realised by the accomplishment of a system's task in the application domain and its incurred costs. There are a number of reasons why this should be the case. The conception from which the concepts originate is an engineering one (Dowell and Long, 1988b). It is new and under development, and its engineering orientation would not necessarily be acceptable to all contributors. In any case, the conception had not been proposed at the time the work reported in the examples was carried out. Further, some contributors have used alternative conceptions, while the conceptions of others are only implicit.

In terms of the conception, the examples are generally less concerned about changes in the applications domain brought about by the work system and more concerned with changes in the computer brought about by the user. Likewise, they are less concerned about the performance of the system and more concerned with the behaviour of the user and ways of optimising that behaviour. Optimisation is typically expressed in terms of general indices such as speed and errors; and not in terms of indices indicating whether or not some specific desired performance for a system has been attained. Further, the knowledge required by the user to operate the computer tends to be specified in terms of the user's mental representation of the device and its operations, rather than in terms of the processes which use the representations. The aim underlying many of the contributions seems to be to identify those mental representations of the user, with respect to the computer and its software, that produce behaviours associated with greater speed and fewer errors. Greater speed and fewer errors may be associated with optimal behaviours, which in turn may be associated with greater effectiveness. However, greater effectiveness does not necessarily equal desired system performance.

The contributions offer many different approaches to optimising user behaviour and many different ways of supporting the knowledge representations required for optimal behaviour. Campion (Chapter 2), for example, assumes that the tactical knowledge required by a naval submarine commander can be supported by using colour to enhance displays depicting the tactical situation. He suggests that colour, by reducing perceived clutter on the screen, and so, for example, making critical aspects of hostile vessel formations easier to interpret, may facilitate the commander's use of mental representations to formulate and execute tactical plans. Reducing perceived clutter may improve performance, increasing the quality of the task product and also reducing the (mental) production costs of the commander. To aid the appropriate assignment of colours to display contents, Campion proposes a structure for the assignment and the procedure necessary to implement the assignment in a system prototype.

An alternative approach is offered by Payne (Chapter 5). His contribution assumes that mental representations supporting optimal behaviour can be facilitated in two ways. Users can be provided with a computer which is easy to learn. That is, which requires little or no new knowledge for its operation, or only knowledge which is easy to acquire. Alternatively, users can be encouraged to learn to operate the computer – in his example a mouse-driven text editor – in ways which support optimal behaviour. As an illustration, knowledge in the form of a minimal conceptual model (that is, representation) of the text editor can support operations like 'move' and 'copy'. However, an elaborated model, with a more complete representation of the consistency underlying the editor, is better able to support a range of 'delete' operations. Payne proposes a notation for the expression of task-action grammars which model the representations users require to carry out computer operations. The notation is applied to the text editor. The notation can be used to reason about the learnability of alternative computer designs and about the ability of different types of user's conceptual model (or representation) to support optimal behaviour. The notation can serve as a system development aid for design or training.

A final example is provided by Whitefield (Chapter 3). He assumes that optimal user behaviour can be encouraged if the computer and its program appropriately support the knowledge (in the form of mental representations) recruited by the user to carry out the actions required by the task. Whitefield reports the development of a model of engineering design activity, as might be involved in the design of a casing for a television monitor or a keyboard. The model expresses the knowledge – of materials, fixings, costs, etc – used to produce a design for the casing. The knowledge is specified in terms of knowledge 'sources' (or representations) interacting via an hierarchic data structure understood to describe designers' behaviour. The model is intended for use by developers of computer-aided design systems to reason about alternative designs for a system. The model helps to ensure that the knowledge used by designers – knowledge of materials, fixings, costs, etc for casings – is supported by the computer. That is, the computer facilitates the recruitment of the appropriate mental representations. Such facilitation in turn encourages optimal design behaviour.

Taken together, the examples provided by the contributions employ many of the concepts described in the framework. The concepts include both those general to human-computer interaction, such as system, behaviour and interface, and those more particular to cognitive ergonomics, such as knowledge, representation and process. The use of some concepts appears similar across contributions, for example, the expression of user knowledge in terms of representations. Other concepts are used differently, for example 'system' to refer to the computer system rather than to the work system (which includes both the user and the device). In general, the contributions concern themselves with the mental representations required to support optimal user behaviour, as reflected by measures of speed and error.

1.5.2   *Methods*

Consistent with the expression of the general problem of human-computer interaction proposed earlier, the methods of human-computer interaction are employed to develop implementable specifications of a desired performance of work. The framework for methods proposed here recursively and selectively uses the concepts for human-computer interaction proposed by Dowell and Long (1988b). In addition, the framework draws on a definition of methods suggested by Colbert, Long and Green (in press).

In the framework, a method is defined as the implementable specification of *behaviour* (or a set of behaviours) necessary and sufficient for the *user* of the method with the aid of *instruments/tools* to perform a *task* (or a set of tasks) with a *goal* in an *organisational context* to a level of *performance* expressed by a set of *criteria*. The behaviours are conceived as the *process* of the method. The task within the method here is one of solving, or supporting the solution of, the general human-computer interaction problem. The outcome of the task is conceived as the *product* of the method. The specification of a method expresses at some *level* of description the *representation* of the *control knowledge* required for that behaviour to accomplish the task. The implementation of the method involves the instantiation of the behaviour which the control knowledge supports and which effects accomplishment of the task. In summary, then, the framework distinguishes: the users of a method (that is, who carries it out); the processes of a method (that is, how it is carried out); and the product of a method (that is, what results from the carrying out).

The methods used by cognitive ergonomics are those whose product is specifications of knowledge. That is, the representations and processes required to support the user's behaviour, such that, interacting, the user and the computer achieve desired performance. Methods involving the complex representations of knowledge and its associated processes may be for use in cognitive ergonomic research. Simplified representations may be for use in cognitive ergonomic practice. Methods vary in the formality with which they are represented (Long and Dowell, 1988), and the extent to which they call upon analytic and empirical behaviours (Long and Whitefield, 1986).The product of a method may contribute more or less directly to specifying the knowledge which supports the user's interaction behaviour. A direct contribution, that is to practice, may take the form of a training programme or the specification of the interface which embodies or implies the knowledge required by the interaction. An indirect contribution, that is developed by research, may take the form of a model representing the knowledge, which is used by someone other than the developer of the model, to create a training programme or specify an interface. The process of a method may be implicit, that is its specification might only be abstracted from a description of its implementation – how the particular training programme was developed or the particular interface designed. The process may also be explicit, that is a complete enumeration of the decisions, actions and their sequencing required to develop a training programme or to specify an interface.

Examples of methods Many of the methods currently practised in cognitive ergonomics are reflected in the contributions reported here. Indeed, all the contributors can be considered to be users of cognitive ergonomic methods in one form or another. Methods for eliciting knowledge include interview methods, for example: Buckley (Chapter 6) identifies the ways in which designers of interactive videotex applications, such as teleshopping, specify user knowledge implicitly as part of their design decisions; and Campion (Chapter 2) identifies the knowledge used by expert operators to compile and edit visually displayed tactical plans supporting a submarine commander's decision-making. Observational methods are also used to elicit knowledge, since knowledge can be inferred from observable behaviour, as illustrated by Whitefield (Chapter 3) in his characterisation of the knowledge used by engineers to design casings. Experimental methods are also used to assess the support afforded by different classes of knowledge – in the case of Barnard et al's work (Chapter 4) different contraction rules for command names – to the behaviour required by the interaction.

In contrast to eliciting knowledge, other methods are used to embody knowledge in the specifications of tasks for implementation in an interaction. For example, Bornat and Thimbleby (Chapter 8) present the context and design decisions underlying a particular interactive text-editor which they developed. The method and principles used to design the text-editor, such as 'simple operations should be simple and the complex possible', may presuppose the knowledge the user brings to the interaction with the text-editor. (That is, knowledge sufficient to support the behaviour demanded by simple operations and sufficient to constitute the basis for the acquisition of additional knowledge demanded by complex operations.) Alternatively, the method and principles may presuppose the conditions under which the extant knowledge can be used and additional knowledge can be acquired, for example, 'fast interaction' or 'edit a picture of the text'. Likewise, Diaper and Johnson (Chapter 7) use methods to embody the knowledge, required by young trainees in microelectronics, computing and automated office applications, in syllabi to support training programmes.

All contributions use cognitive ergonomic methods, either for eliciting or embodying the knowledge necessary to support the behaviour required by an interaction. However, some contributions also attempt to develop the methods. Campion (Chapter 2), for example, proposes a method for assigning colour to tactical visual displays to facilitate the use of the submarine commander's tactical knowledge. Barnard et al. (Chapter 4) report a method for the development of a science base for the naming of computer commands. Finally, Payne (Chapter 5) describes a method for reasoning about the additional knowledge which alternative computer designs require for successful interaction with the device.

Of the contributions reporting the development of cognitive ergonomic methods, many exhibit differences concerning the intended users of the methods, as well as concerning the process and the product of the methods. Concerning intended users, some contributions target human factors engineers, for example, the colour assignment method of Campion (Chapter 2). Others target software engineers, for example, the engineering design task model of Whitefield (Chapter 3). Methods reported by other contributors are aimed more generally at interaction developers, typically designers. For example, information technology training programme designers in the case of Diaper and Johnson (Chapter 7), and videotex display designers in the case of Buckley (Chapter 6).

Concerning the process and product of cognitive ergonomic methods, some contributors concentrate their efforts on specifying the product of a method rather than the process of its use. For example, the product of Whitefield's work (Chapter 3) is a model of the engineering design task expressed as an hierarchical data structure (the 'blackboard') and sets of knowledge sources which read from and write to the blackboard. The product of Payne's work (Chapter 5) is a notation for expressing task-action grammars, which model users' competence. Both products are intended to form part of a method for reasoning and deciding about alternative forms of interactive computer design. So far the product, in both cases, is more developed than the process necessary to use it. In contrast, other contributors report both the product and the process for its use. For example, Campion (Chapter 2) presents both a structure for colour assignment (the 'display separation description framework') and the procedure for its use. Likewise, Diaper and Johnson (Chapter 7) offer a representation for training syllabi (the 'knowledge representation grammar') and the procedure for its use to express particular syllabi and to transform them into training programmes.

Finally, contributors differ as to the product itself. Only one contribution, that of Bornat and Thimbleby (Chapter 8), specifies (and implements) the interaction and the associated interface (although Diaper and Johnson (Chapter 7) specify the knowledge in terms of training syllabi). Most other contributions are proposed as aids to the production of implementable specifications. Whitefield's task model (Chapter 3), Payne's task-action grammar (Chapter 5), Barnard et al's candidate scientific principles (Chapter 4), etc. are all intended to support the practice of interactive computer systems development.

In summary, then, although all contributions use cognitive ergonomic methods, only some contributions concern the development of such methods. Of these, some concentrate on the product of the method, while others are concerned with both the product and the process. Although the products may be expressed formally – or at least explicitly – as notations etc., the processes are often not so expressed. In addition, contributions are not explicit about the organisational contexts, criteria for performance etc. and other features of methods identified by the framework. The underspecification in the contributions of the processes of the methods, and certain of their other characteristics, result from their early stage of development. Greater completeness would be expected to accompany their further development. Lastly, intended users of cognitive ergonomic methods may be either human factors engineers, software engineers or more generally designers. The product of the methods may contribute directly or indirectly to the development of implementable specifications of some desired performance of work.

1.5.3   *Interaction development practice*

Interaction development practice (conventionally termed system development) is none other than the process required to accomplish the task of solving the general human-computer interaction problem. That is, to develop specifications of computer behaviour and user behaviour, such that, once implemented, a desired performance of work results from their interaction. The product of the practice is the developed interaction, as embodied in a particular system. The concepts and methods of human-computer interaction, then, as described earlier, provide, with some additions, a sufficient framework for modelling the practice of interaction development expressed as a process. The framework draws on descriptions of system development to be found in Long (1986) and Walsh, Lim, Long and Carver (1988).

According to the framework, the aim of human-computer interaction is to develop implementable specifications of human-computer interactions for a desired performance of work. It follows, then, that any process of interaction development practice must at least subsume the activities of specification and implementation. The activity of *specification* produces a structural and behavioural representation of the system to support the interaction. The activity of *implementation* instantiates the specification in a particular system which supports the interaction. Further, since interactions are for a desired performance of work, it follows that practice must subsume two further activities – performance setting and performance evaluation. *Setting* of performance (akin to the more conventional 'user requirements' identification) indicates the tasks to be carried out by the system and the quality of performance to be achieved by the product of the task. The product of setting may ideally be expressed in the form of (performance) criteria. *Evaluation* of performance applies the criteria to the implemented system to assess its quality of performance. Performance setting typically precedes specification and may be conceived as part of the specification process. Performance evaluation typically follows implementation and may be conceived as part of the implementation process.

These categories of practice (performance setting; specification; implementation; and performance evaluation) provide the flexibility required to accommodate variations in the process of interaction development practice reflected in the examples of cognitive ergonomics reported here. Other activities may be added, for example, system installation and maintenance, but these may be subsumed under the category of implementation. Activities may assume more than one form of practice. For example, fast prototyping and formal decomposition are alternative ways of producing a specification. Activities may also enter into different relations with one another. For example, performance may be evaluated on the basis of a system's specification rather than its implementation. Lastly, activities may be carried out by different interaction developers. For example, performance setting may be carried out by human factors developers or software developers (that is systems analysts) or both.

The contribution of cognitive ergonomics to interaction development practice is to exploit its concepts and methods to specify the knowledge supporting user behaviour in the interaction (as required by the practice of performance setting, specification, implementation and performance evaluation). Suppose, for example, performance evaluation of a specific system assesses current performance to be inferior to that desired, as expressed by performance criteria. Cognitive ergonomic interaction development practice would suggest changes to the current specification of knowledge (typically expressed in the form of representations and processes). The associated change in user behaviour would be expected to reduce the differential between current and desired system performance. The substitution of existing domain object labels for computer jargon labels might be such a change. If the user's existing knowledge, expressed as the representation of the object labels and their associated processes, optimises interaction behaviour, the system is more likely to attain desired performance.

Examples of interaction development practice The examples reported here provide few illustrations of cognitive ergonomic interaction development practice. Only the contributions of Diaper and Johnson (Chapter 7), and Bornat and Thimbleby (Chapter 8), both specify and implement the knowledge to support user behaviour. It is true that Buckley (Chapter 6) reports the evaluation of a specific videotex system ('Prestel'), and that Payne

(Chapter 5) reports the analysis of a particular text editing system ('Te-dit'). However, both are primarily concerned with generic interaction development. Payne proposes a notation for system analysis which expresses the knowledge needed by the user to generate the behaviour required by the interaction. Buckley's identification of the problems and difficulties, experienced by users attempting to use videotex services for transaction processing, is used to explore ways in which the practices of videotex frame designers might be influenced. Other examples, however, do not involve the analysis of particular instances of interaction development. Campion's structure and procedure for colour assignment (Chapter 2), Whitefield's knowledge-based model of mechanical engineering design (Chapter 3) and Barnard et al's principles of contraction rules for command names (Chapter 4), are all intended to have a wider scope of application than a particular system. They do not, therefore, constitute instances of cognitive ergonomic interaction development practice.

However, even examples with primarily generic concerns make assumptions about the interaction development practice to which their indirect contributions are aimed. There is little reference to performance setting, even in terms of user requirements identification. Nor is there much reference to performance evaluation, although reference is frequently made to optimal behaviour (for essentially the same point, see Section 1.5.1). In contrast, there is frequent reference to design (subsumed here under specification) and system development (subsumed under interaction development). Campion's procedure for colour assignment (Chapter 2) is intended to contribute indirectly to the practice of prototype design. Likewise, Whitefield's task model of engineering design (Chapter 3) and Barnard et al's principles when related to command contraction (Chapter 4) are intended to contribute to design activity by specifying the knowledge required to induce optimal behaviour. Most examples, then, are intended to provide an indirect rather than a direct contribution to the specification activity of the practice of cognitive ergonomic interaction development. (Indirect contributions in the form of support for practice are considered in Section 1.5.4 which follows.)

Diaper and Johnson (Chapter 7), and Bornat and Thimbleby (Chapter 8), in contrast to the other work, constitute examples of cognitive ergonomic practice which contribute directly to interaction development practice (although both also have generic concerns). Diaper and Johnson specify knowledge in the form of syllabi which, when embodied in training programmes, support the behaviour required to operate information technology devices, such as word processors. Their contribution exemplifies versions of performance setting, specification, and implementation. (But not, as they realise, performance evaluation. Evaluation would have required: the syllabi to be embodied in particular training programmes; the programmes used to train information technology naive students on particular devices; and the assessment of the performance of the systems, whose devices the trainees operated with respect to desired performance.)

Bornat and Thimbleby describe how they developed the text-editor 'Ded'. They refer to tasks that can be achieved by the user with the editor, such as writing letters and editing computer programs. They cite criteria for performance or at least for optimal behaviour, for example, that the editor can be used straight away. In specifying and implementing the editor, they also (at least implicitly) specify and implement the knowledge required to support user behaviour in the interaction. Some of the knowledge required is dictated by the interface itself, for example, the layout of a keypad (which might or might not be labelled). Other knowledge is specified in the manual, for example, the use of advanced facilities. Lastly, they assess performance, or at least optimal behaviour, in terms of the time taken by people to use the editor successfully. Bornat and Thimbleby, then, illustrate all the categories of interaction development practice identified in the framework, albeit informally.

This poses an apparent paradox. The most complete illustration of cognitive ergonomic interaction development practice is provided by two computer scientists. In addition, Bornat and Thimbleby apparently make little use of the kind of indirect cognitive ergonomic contributions offered elsewhere in this book, for example, models of the task, notations for reasoning about a program's learnability, etc. This may suggest that the kind of indirect contributions to practice offered by cognitive ergonomics is redundant.

There are a number of reasons for the apparent paradox. First, Bornat and Thimbleby began development of 'Ded' in the late seventies. This was a time when cognitive ergonomics was only beginning to establish itself and its possible contribution to development practice. Little explicit or formal cognitive ergonomic support, then, was available. Second, the need for a cognitive ergonomic contribution is highlighted by the development of the text-editor. For example, the principle of 'modelessness' advanced by the authors might reasonably suppose that having to change modes increases the user's cognitive processing load, because, for example, a representation of the current mode needs to be kept in memory. However, if 'modelessness' requires the selection among many options, the load associated with searching the options might be greater than that associated with remembering the mode. Cognitive ergonomics attempts to address such issues. Third, some of the design principles advanced are similar to principles contributed by cognitive ergonomics, for example, 'always show something is happening' (on the screen) is like the cognitive ergonomic principle of 'provide feedback for every user action'. Lastly, Bornat and Thimbleby belonged to the target user group for which 'Ded' was developed. They were frequent and expert users of 'Ded' as well as computer scientists/designers/programmers. Their insights into their own and their colleagues' cognitive requirements (albeit implicit and informal) might be expected to be greater than their insights into the cognitive requirements of casing engineers (had they attempted to develop a computer-aided design system), or of submarine commanders (had they attempted to develop a colour graphics tactical picture system). Whitefield's task model (Chapter 3) might be expected to offer greater insight in the former case, and Campion's colour assignment procedure (Chapter 2) in the latter – both sets of insights having potential for contributing indirectly to interaction development practice.

The paradox that Bornat and Thimbleby's work should provide the most complete illustration of the practice of cognitive ergonomic interaction development, then, is only apparent. Further, the apparent redundancy of cognitive ergonomics with respect to the specification of the knowledge required to use their text-editor may have derived from particularities of the design, the designers and the time at which the design was undertaken.

Taken together, the work reported here provides few examples of the practice of cognitive ergonomic interaction development. Most of the work supposes its contribution to be indirect. In addition, most of the work supposes its indirect contribution to be to the practice of system specification (more conventionally design).

1.5.4   Support for interaction development practice

Support for interaction development practice is none other than the knowledge of the discipline of human-computer interaction which indirectly contributes to the solution of the general human-computer interaction design problem. Support takes the form of principles which, when applied by practice, ensure that specifications of computer behaviour and user behaviour, when implemented, interact to produce a desired performance of work. Principles pertain both to the process – that is how the general design problem is solved, and to the product – that is the properties of the interaction which constitutes a solution to the problem. Principles of both kinds are the product of support activities. The performers of the activities are the supporters of interaction development practice. The concepts and methods of human-computer interaction, then, as described earlier provide, with some additions, a sufficient framework with which to model the support for interaction development practice expressed as a process. The framework draws on a description of support activities to be found in Long (1986) and Long and Dowell (1988).

In the framework, the *real world* – that is the world as an object of intent (see the concept of work defined earlier in Section 1.5.1) – is contrasted with a *representational world* whose function is to support intended change in the real world. These two worlds may be related by means of *intermediary representations* and activities which transform one representation into another. The representations are the products of the activity processes. In the figures which follow, representations are shown as boxes and transformational activities as directional arrows. Figure 1.1 shows the activities of interaction development practice which are to be supported, modelled in terms of the framework.

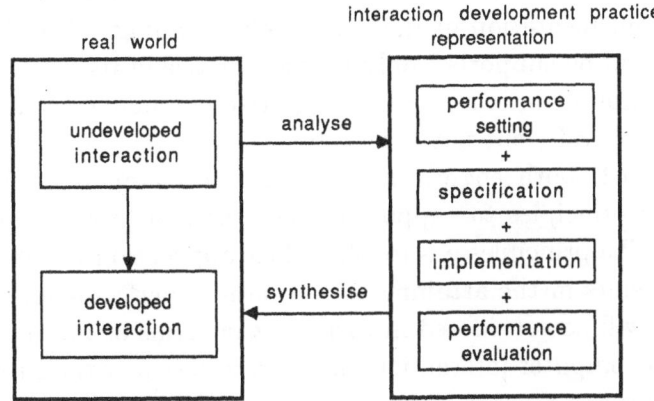

Figure 1.1: Model of Human-Computer Interaction development practice

The real world of interaction development, as illustrated in Figure 1.1, shows the intentional change of an undeveloped interaction into a developed interaction. The undeveloped interaction might be either the absence of an interaction, an extant interaction not involving a computer, or the new development of an existing user-computer interaction. Analysis of the undeveloped interaction produces a set of representations reflecting the interaction development activities described earlier in Section 1.5.3 (that is, performance setting, specification, implementation and performance evaluation). Synthesis of the interaction development representations with the real world produces the developed interaction. The criterion for the success of the developed interaction is whether it corresponds to some intended change in the real world. That is a change related to a new desired performance of work.

Support for interaction development practice is of three types – practice, engineering and science. The practice of interaction development supports itself in a manner similar to that of a craft which is supported only by its own practice. The practice of performance setting, specification, implementation and performance evaluation leads the practitioners to acquire and maintain informal knowledge sufficient to support subsequent practice. For example, the use of colour to code graphical displays is part of current interaction development practice. The criterion is that colour coding works or seems to work. However, since the knowledge is informal, it embodies no principles of colour coding able to guarantee that an implementable specification including colour will meet a desired performance of work. Without principles, this informal knowledge cannot be applied generally, that is the success of colour coding in supporting one interaction will not necessarily be followed by success in supporting another. Neither can the informal knowledge be validated, other than by the success or failure of each implementation of the use of colour coding. Were interaction development to be supported only by the practice of its own activities, both human-computer interaction and cognitive ergonomics could lay claim to being only a craft, but not a discipline.

In contrast, both engineering and science claim to be disciplines and to provide principles to support the activities of interaction development practice. The principles ensure that their application to interaction development results in the attainment of a desired performance of work. Each discipline will be considered in turn, first in terms of the support provided by human-computer interaction and then in terms of the support provided by cognitive ergonomics.

Engineering is concerned generally with the production of artefacts such as buildings (civil engineering), plant (mechanical engineering), electrical appliances (electronic and electrical engineering), etc. Engineering practice seeks solutions to the specific design problems posed by the need for individual artefacts such as buildings, plant, electrical appliances, etc. Engineers specify the artefacts whose implementation will meet the need. In contrast, engineering research develops the principles, both for the properties of the artefacts and for the methods by which they are produced. The principles ensure that the artefacts, when implemented, meet their specification.

The support provided by human-computer interaction engineering research to interaction development practice is shown in Figure 1.2, modelled in terms of the framework. The real world of human-computer interaction engineering research is the interaction development practice illustrated in Figure 1.1. The role of engineering research is to increase the number and improve the generality and power of principles available to interaction developers. Analysis of development activities produces a set of representations reflecting the current state of interaction development principles. In the model, the principles are identified as human factors engineering principles, software engineering principles and other relevant engineering principles. The representations are used to establish new principles of these three kinds. Synthesis of the engineering research representations with the real world of practice produces developed interactions whose implementation meets their specification. The criterion for the effectiveness of human-computer interaction engineering research is that the principles it develops ensure the correspondence between the specification of an interaction and its implementation. That is, the principles work. This criterion of effectiveness constitutes one form of validation for principles (for a second form – validation by explanation – see later in this section).

The contribution of cognitive ergonomics to human-computer interaction engineering research is to establish human factors engineering principles whose application ensures that the implementation of the knowledge supporting user behaviour in the interaction meets its specification. Suppose some desired level of performance for a submarine command and control system was expressed as the identification of all formations of enemy surface vessels (see Chapter 2 by Campion). Further, suppose colour coding was a proven principle of cognitive ergonomics for reducing perceived clutter on graphical displays (and so for facilitating the use of the submarine commander's knowledge to support behaviour expressing tactical reasoning). Then incorporating the principle into the command and control system's interaction would contribute to the system exhibiting desired performance (to the extent that tactical reasoning with the aid of colour determines identification). The principle of colour coding is here expressed as a property of

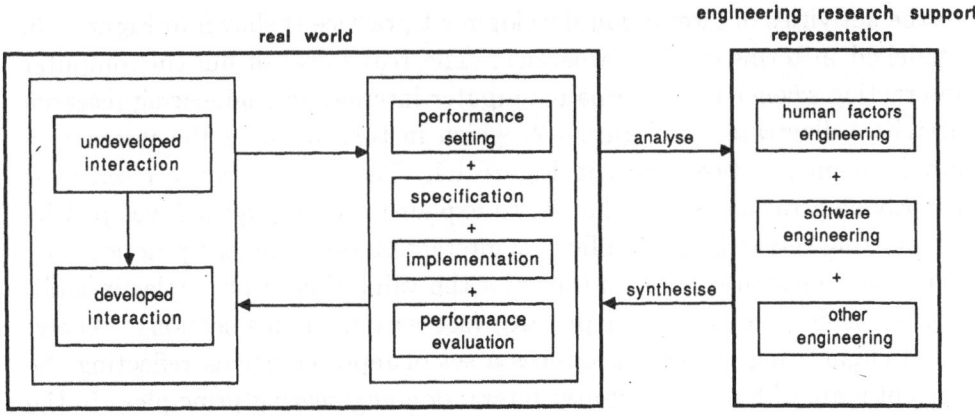

Figure 1.2: Model of Human-Computer Interaction engineering research support for interaction development practice

the computer (that is the graphical display) and of the user's interaction (that is tactical reasoning). In addition, the method for assigning colours to displays would also constitute a principle, if it ensured an implementation met its specification. Cognitive ergonomic engineering research, then, as part of human factors, aims to provide principles to support interaction development practice.

Science is concerned generally with the construction and validation of laws and theories to explain and to predict phenomena. The laws and theories (including models), when operationalised in terms of the phenomena, constitute truths concerning the world and are the criteria against which science is judged. Science is a discipline whose knowledge – in terms of the phenomena, their measurement and the theories to explain them – is required to be explicit and objective.

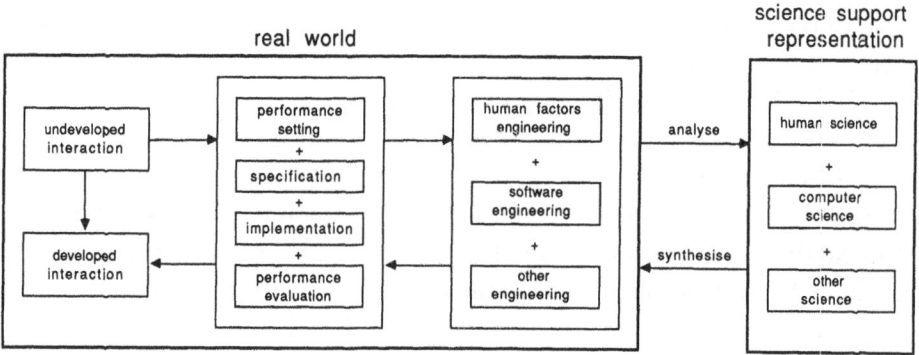

Figure 1.3: Model of Human-Computer Interaction science research support for engineering research and hence for interaction development practice

The support provided by human-computer interaction science research to the activities of interaction development practice is shown in Figure 1.3, modelled in terms of the framework. The real world of human-computer interaction science is the human-computer interaction engineering research activities illustrated in Figure 1.2, which in turn includes the interaction development practice shown in Figure 1.1. The role of science research is to provide a rationale for (that is, to explain) the principles developed by engineering research and applied by interaction development practice. Explanation constitutes a second form of the validation of principles in addition to their effectiveness in practice (see earlier in this section). Analysis of engineering research produces a set of representations reflecting the current state of human-computer interaction engineering principles. In the model, the science representations are identified as those of human science, computer science and other relevant sciences. The representations are used to explain, and so validate, the engineering research principles. Synthesis of the science representations with the real world of engineering research results in explicated, and so validated, engineering principles. When applied by interaction development practice, these principles produce interactions whose implementation meets their specification. The criterion for the effectiveness of science support for interaction development practice is that its theories are true when expressed as propositions about the world. Further, that they explain, and so validate, the principles of engineering research.

The contribution of cognitive ergonomics to the human science support for human-computer interaction engineering research is to establish theories of human behaviour. These ensure that human factors engineering principles, when applied to developing interactions, produce implementations of the knowledge supporting user behaviour which meet their specification. Suppose, as earlier, that colour coding is a cognitive ergonomic principle for reducing perceived clutter on graphical displays. A scientific theory supporting the principle would have to describe and to explain the phenomena whereby colour coding facilitates the use of the submarine commander's knowledge to support tactical reasoning behaviour. The explanation would have to invoke the representations and processes of perceptual and conceptual knowledge, and their interaction, with the applications domain, and the representations thereof provided by the graphical display. The theory thus explains, that is rationalises, and so validates, the specification embodied in the principle. In so doing, it ensures that the implementation of the principle meets this specification, that is, ensures the principle is effective. Cognitive ergonomics, then, as part of the human sciences explains, and so validates, principles of human factors engineering research.

Examples of support for interaction development practice Most of the present cognitive ergonomic examples may be understood, within the terms of the framework, to concern the development of principles to support interaction development practice. As such, they constitute examples of human-computer interaction engineering research support and, in particular, of human factors engineering research support (rather than of either interaction development practice or science support).

Campion (Chapter 2), for example, is attempting to develop principles for the colour coding of displays and for a procedure to assign colours to the representations which are displayed. He suggests that the principle of colour coding may be used to effect the perceptual decluttering of tactical displays. He also proposes a levels principle for assigning colour to displayed entities. Payne (Chapter 5), in contrast, is attempting to develop principles for users' conceptual models and their use. He proposes

a notation for the models which is able to express the consistency underlying an interaction, and so can be used to reason about the learnability
of alternative designs. Similarly, Whitefield (Chapter 3) is attempting to
develop principles for task models and their use. He proposes a structure
which is able to represent the knowledge required for performance of the
user's tasks, and so can be used to reason about the support provided by
alternative designs.

In contrast, the contribution of Barnard et al. (Chapter 4) seeks to
demonstrate how the science base representation can be developed with
a view to the indirect support of interaction development practice. They
report a number of empirical regularities from experiments on the naming of computer commands and suggest some candidate principles, such
as truncation, which can both support development practice and be explained, and so validated, by theory.

Since Diaper and Johnson (Chapter 7), and Bornat and Thimbleby
(Chapter 8) illustrate interaction development practice (see Section 1.5.3),
it is possible to consider whether they used any of the three types of support identified earlier. Both contributions, in fact, appear to have exploited
support – but of different kinds. Diaper and Johnson develop the principle
of levels of description and apply it to structure the training syllabi. The
principle is derived from a hypothetical science representation of a structure relating knowledge to human behaviour. Bornat and Thimbleby also
develop and apply principles in the development of the text editor, for example, 'modelessness'. However, the latter principles, as cognitive ergonomics
engineering principles (as opposed perhaps to software engineering principles) are based on interaction development practice.

Taken together, then, most of the present contributions are concerned
with the development of human factors engineering principles for the support of interaction development practice. All of the principles are under
development, and most are at an early stage. However, the support provided by practice, and that by the science base are also illustrated, as well
as the possible relationship between the science base and engineering principles.

1.5.5   *Framework summary*

The aim of this section has been to introduce the contributions
of cognitive ergonomics to human-computer interaction within a coherent
perspective. Coherence has been provided by a framework with which to
model the different kinds of contribution. The framework consists of concepts, methods, interaction development practice and support for interaction development practice. The framework has been used to characterise

the different contributions.

Concerning the examples of concepts, there is much overlap between contributions, for example, knowledge expressed as representations and processes. There are, however, also important differences – for example, the technology system (the computer) as opposed to the work system (the user and computer). In general, the contributions are concerned with the knowledge required to support optimal human behaviour, as reflected by measures of speed and error (rather than optimal human behaviour as it relates to optimal computer behaviour for desired work performance). Concerning the application of methods, a number of contributions report the development of both formal and informal products and processes. However, perhaps due to their early stage of development, many of the methods are underspecified, for example, with respect to organisational contexts, criteria for performance, etc. Concerning interaction development practice, almost all contributions are indirect, and so constitute activities of support, which suppose their target to be system specification (that is design). Concerning support for interaction development practice, most contributions are attempting to generate human factors engineering research principles. The principles are at an early stage.

In summary, then, the contributions show cognitive ergonomics to be active both concerning the development of concepts and methods, and concerning engineering research support for practice. The practice of cognitive ergonomic interaction development is not well represented by the contributions. The various developments, however, are at an early stage and this reflects the current state of the field.

1.6 Organization of the book

The secondary aim of this chapter is to describe how part of the framework has been used to organise the contributions within the book. Before identifying which aspects of the framework are involved, consider why the complete framework has not been used. First, the framework itself continues to be under development, and was not available when the contributions were begun. Second, given the differences within the field of cognitive ergonomics (see, for example, Section 1.5.1), there would not have been agreement among the authors concerning its status and use, in particular with respect to its engineering orientation. Third, the framework might have been overly complex as an organising scheme.

The part of the framework distributed to contributors and used to organise the book is shown in Figure 1.4. The model should be considered as a summary of the model in Figure 1.3 (although historically it provided the basis for the elaborated model). The model is of human-computer interaction science support of which cognitive ergonomics, as it relates to human science, is a part. For the purpose of organising the book, engineering research should be considered part of science support activities and human factors engineering research as part of human science. The distinction between science and engineering shown in Figure 1.3 is omitted in Figure 1.4 for the sake of simplicity. The real world of science (and engineering) support consists of people, computers and other entities, such as organisations. (The undeveloped and developed interactions of Figure 1.3 are between these people and their computers.) Scientific (and engineering research) knowledge to support changes in the real world, that is the development of new computerised work systems, is shown as a set of science representations which include human science, computer science and other science. Scientific (and engineering) knowledge is acquired by two transformations. The activity of *analysing* the real world produces an *acquisition representation* which typically supports laboratory simulation for experimentation. The activity of *generalizing* the findings from experiments produces the engineering principles and scientific theories of the *science knowledge representation*. Scientific knowledge is applied likewise by two transformations. The activity of *particularising* the science representation produces an *application representation*, typically in the form of guidelines or principles for the design of tasks, technology or the working environment. The activity of *synthesising* this activity with the real world of people and computers contributes to the production of new or changed computerised systems of work, that is, interaction development practice.

In terms of the framework, human-computer interaction generally, and cognitive ergonomics in particular, consists of two main activities. One acquires knowledge about people and computers – the transformations of analysis and generalisation in Figure 1.4. The other applies knowledge – the

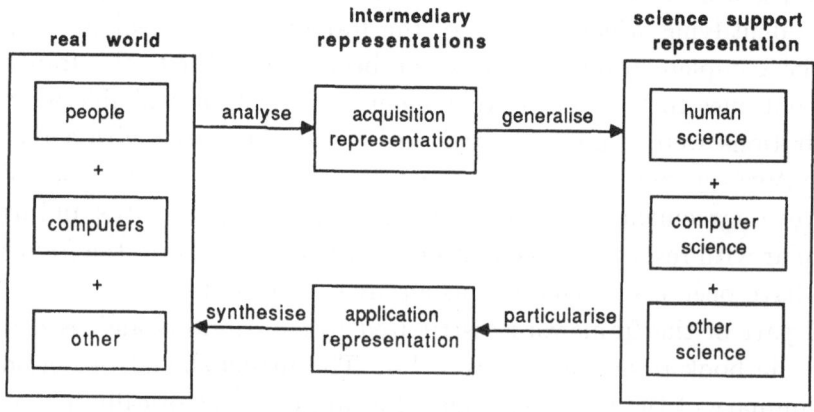

Figure 1.4: A model of Human-Computer Interaction science support

transformations of particularisation and synthesis also shown in Figure 1.4.

The framework will now be used to characterise the contributions, in particular with respect to acquisition and application activities. The characterisation will serve to introduce the reader to each contribution. By relating each example to this common framework, the relations between contributions will be exposed. The framework, then, is used to feed forward to the contributions. In addition, each chapter indicates its relation to the framework and so feeds back to the present introduction. Taken together, the forwards and backwards referencing should provide the book with sufficient coherence to support understanding of the contributions and their relations.

In Chapter 2, Campion describes a cognitive approach to research on the assignment and assessment of colour in computer generated displays in submarine command and control systems. The aim is to generate knowledge of how colour can be used to support the tactical reasoning of the commander. The information is intended to be both theoretical (that is generalisable) and subject to empirical test. The goal is to enable the prototyping of new naval systems. Campion's contribution to cognitive ergonomics is to suggest a method by which the tactical knowledge required by a submarine commander can be specified with respect to the use of colour to enhance displays depicting the tactical situation.

In terms of the model of human-computer interaction science support shown in Figure 1.4, the chapter primarily illustrates the construction of an acquisition representation. Observation and analysis of submarine command and control system interactions, including identification of the knowledge supporting them, resulted in a model of command and control, expressed as a tactical system – an initial acquisition representation. The model was then re-expressed in the form of instructions constituting a simulated laboratory task to assess alternative colour assignments (a second acquisition representation). The chapter also describes the data derived from the experiments. The data are in conjunction with criteria for optimisation used to suggest that colour, by reducing perceived clutter on visual display screens (and so for example facilitating the interpretation of critical aspects of hostile vessels), may support the commander's use of mental representations to formulate and execute tactical plans. The resulting method is summarised as a procedure for colour assignment which constitutes an application representation to aid in the prototyping of new submarine command systems.

In Chapter 3, Whitefield describes a case study in which a model of mechanical engineering design activities was constructed. The model is intended to be applied by developers of computer-aided design systems to support the production of suitably structured programs. A representation from artificial intelligence provided the model's analytic structure and an observational study its empirical content. The model was subsequently used to reason about the consequences for the quality of design of alternative program structures, assessed in a simulated design task. Whitefield's contribution to cognitive ergonomics is to specify the knowledge used by designers, for example, knowledge about materials, fixings, costs etc, in the design of artefacts such as television casings, in a form which can be used to reason about the alternative support a computer might provide.

In terms of the model shown in Figure 1.4, analysis of extant engineering design task interactions identified the acquisition representation which formed the basis for the empirical study. The data from the study when expressed within the analytic structure formed a model of design knowledge (the 'blackboard' model) which constituted an initial application representation. Unlike the model of human-computer interaction science support, Whitefield's approach is to transform the acquisition representation of engineering design into an application representation without the additional transformations which would have been required to gain support from the science representation. Whitefield considers this more direct route from knowledge acquisition to knowledge application to be an engineering approach to cognitive ergonomics. This view can be considered in the light of the models of human-computer interaction engineering and science support for interaction development practice shown in Figures 1.2 and 1.3.

In Chapter 4, Barnard et al. describe how laboratory experimentation can be used iteratively to establish empirical regularities for interpretation in terms of the science representation. The chapter explores, in the context of a document processing task, issues concerning the principles which could be recruited to support the naming of computer commands in a new or modified system. The issues include: the different namesets which could be used; how novel meanings might be learned; and alternative schemes for command name abbreviations. The resulting empirical regularities are used to formulate principles characterising user-computer interactions related to command names. Principles such as truncation, they argue, can both support interaction development practice and be explained by theory. Barnard et al's contribution to cognitive ergonomics is to attempt to formulate the principles which relate the knowledge representations, recruited by the naming of commands in the interactions, to the regularities in behaviour which result.

In terms of the model shown in Figure 1.4, Barnard et al. exploit an acquisition representation of command naming in document processing to use laboratory experiments for gathering data to explore potential regularities in interactive behaviour. The regularities are generalised to form the science representation. The science representation of principles, empirical regularities and theories then becomes available to be expressed as an application representation intended to support document processing interaction development practice. The main emphasis of Barnard et al's chapter – that is, building up the science representation – can be contrasted with the main emphasis of Campion's chapter – that is the derivation of an acquisition representation of the commander's task, and the main emphasis of Whitefield's chapter – that is the direct relationship between the acquisition and the application representations.

In Chapter 5, Payne is concerned with learning and learnability. A 'cognitive architecture' is proposed in the form of a psychological notation. This notation may be applied to any task domain to model competence in terms of the knowledge needed to use the computer for the task. The notation is termed 'task-action grammar'. It is used to analyse alternative task structures for a text editor and their implications for learning. A minimal conceptual model or representation of the text editor is shown to be able to support interactions involving 'move' and 'copy' operations. An elaborated model, with a more complete representation of the consistency underlying the editor, is shown to provide better support for a range of interactions involving 'delete' operations. The contribution of Payne to cognitive ergonomics is to propose a notation able to specify the knowledge required by an interaction. The specification facilitates reasoning about alternative forms of interaction resulting either from different devices, or from different user knowledge about the same device.

In terms of the model shown in Figure 1.4, Payne illustrates how a notation (and in particular a psychological notation) allows the mapping of a science representation onto an application representation (that is particularisation) within a single theory. The psychological theory, within the science representation, expresses cognitive limitations with respect to some task domain. The use of the notation to model competence (the application representation) allows interaction developers to reason about the learnability of alternative designs. The support provided to developers is intended to fulfil the engineering aim of predicting learning and learnability in new interactive systems.

In Chapter 6, Buckley describes work intended to improve the dissemination and applicability of research findings from a project concerned with videotex (in particular their use to support teleshopping). The work of videotex designers was analysed in an attempt to specify the vehicle and form of a representation suitable for supporting design practice to improve usability. Designers were found to learn from the work of other designers rather than to rely on design manuals or research reports. The chapter suggests that a representation compatible with design practice would be one which combines the vehicle of a videotex database with the form of dialogue scenarios. The scenarios would be contrasted in terms of device features sensitive to human behaviour. For example, the finding, that the effectiveness of the type of response frame for ordering goods varies with the type of order, could be demonstrated by implementing the relevant set of videotex dialogues. Designers would be able to recruit the more usable relations between response frames and ordering into their own designs. The contribution of Buckley to cognitive ergonomics is to illustrate how, to be effective, research findings concerning the knowledge required by people to use computers – in this case shopping – need an expression compatible with that of interaction development practice.

In terms of the model shown in Figure 1.4, the chapter illustrates the re-expression or particularisation of candidate research findings for the science representation into an application representation. The proposed representation of videotex dialogue exemplars is intended to support interaction development decisions by encouraging synthesis to occur between research findings and their application.

In Chapter 7, Diaper and Johnson describe the development and application of a method for syllabus design – termed 'Task Analysis for Knowledge Descriptions'. The method involves: the selection, observation and description of tasks; the successive re-expression of the task descriptions – as lists of generic actions and generic objects, then as sentences in a knowledge representation grammar, and finally as a syllabus constructed from the sentences. The supporting psychological theory of human knowledge underlying the method is also described. The syllabus was intended to aid the teaching of information technology skills to young trainees. Diaper and Johnson's contribution to cognitive ergonomics is to have provided a method, its representations and a rationale for specifying knowledge in the form of syllabi. When embodied in training programmes, the syllabi support the behaviour required to interact with information technology devices.

In terms of the model shown in Figure 1.4, the chapter illustrates all the representations and their transformations. The science representation is a psychological model of knowledge expressed as an analytic structure embodying levels of description. The initial task description and its successive transformations constitute acquisition representations and the syllabus itself constitutes an application representation. Synthesising the syllabus with particular training requirements would produce a training programme.

In Chapter 8, Bornat and Thimbleby describe the context and design decisions underlying the development of a text-editor named 'Ded'. An overview of 'Ded' is presented and the historical background of the development characterised. 'Ded' evolved in a particular social environment and illustrates a prototyping approach to design. Various design ideas for the text editor are introduced, for example, 'only one mode', along with a number of design principles, for example, 'always show something is happening'. Problems with the text editor are reviewed and conclusions drawn concerning the design activity which led to the development of 'Ded'. These conclusions include the need for explicit trade-offs in design, the importance of social factors and the success of the evolutionary approach.

Bornat and Thimbleby's contribution to cognitive ergonomics is threefold. First, it illustrates the craft practice or phenomenological cognitive ergonomic activities which may be implicitly and informally involved in interaction development practice. Second, it provides cognitive ergonomics with design ideas and principles. Their consequences for the production of specifications of the knowledge required to use computers need to be made explicit, and their validity need to be established. Last, the contribution identifies the requirement for supporting interaction development practice which the discipline of cognitive ergonomics has to meet, if it is to demonstrate its superiority over the craft approach.

In terms of the model shown in Figure 1.4, the chapter illustrates the phenomenological derivation of an implicit acquisition representation, which informally characterises tasks such as writing letters and editing computer programs. It also illustrates the use of an informally derived application representation – as expressed by design ideas and principles – which supports the synthesis of the new interaction development in the form of 'Ded'. It should be noted, however, that the present description is post hoc and only serves to relate this chapter to the rest of the contributions. As indicated by the authors, the framework, and in particular the part used to organise the book, influenced neither the design of 'Ded', nor the description of its development.

This now completes the summary of the individual chapters, the characterisation of their contribution to cognitive ergonomics and the description of their relationship to the model of human-computer interaction support activities used to organise the book.

1.7   Summary and conclusions

Human-computer interaction as practice and discipline, in spite of considerable current interest and activity, is still in the early stages of its development. The same can be said of cognitive ergonomics. Indeed, it is argued by some that the idea of producing implementable specifications of the knowledge required by humans to use computers is unrealisable in practice (and perhaps in principle). Attempts to do so are, thus, misguided. Interaction development practice by this view is a craft and will remain so.

The general view taken by the contributors to this book is that human-computer interaction and cognitive ergonomics practice are very much craft activities at this time. However, the need and potential for a discipline developing engineering and scientific principles to enhance interaction development practice continue to increase. The contributions are offered in support of this view, and readers are invited to make their own judgments on the progress of the undertaking.

Acknowledgement

This chapter has greatly benefitted from criticism from and discussion with my co-editor Andy Whitefield, and also with John Dowell and Andrew Life – colleagues at the Ergonomics Unit, University College London. Any remaining infelicities of thought or expression are entirely my own.

Colbert, M., Long, J., and Green, D.W. (in press) Methods for the development and application of knowledge-based systems: some features, constraints, and issues. In D. Berry and A. Hart (Eds) *Expert Systems: Human Issues*. London: Kogan Page.

Dowell, J., and Long, J.B. (1988a) Towards a paradigm for human-computer interaction. In E.D. Megaw (Ed) *Contemporary Ergonomics 88*. London: Taylor and Francis.

Dowell, J., and Long, J.B. (1988b) Human-computer interaction engineering. In N. Heaton and M. Sinclair (Eds) *Designing End-User Interfaces*. Oxford: Pergamon Infotech.

Long, J.B. (1986) People and computers: designing for usability. In M.D. Harrison and A.F. Monk (Eds) *People and Computers: Designing for Usability. Proceedings of HCI '86*. Cambridge: Cambridge University Press.

Long, J. (1987) Cognitive Ergonomics and Human-Computer Interaction. In P. Warr (Ed) *Psychology at Work*. Third Edition. Harmondsworth, Middx: Penguin.

Long, J.B., and Dowell, J. (1988) *Formal methods: the broad and the narrow view*. IEE Colloquium on Formal Methods and Human Computer Interaction. February 1988.

Long, J.B., and Whitefield, A.D. (1986) *Evaluating interactive systems*. Tutorial given at HCI '86, University of York, September 1986.

Walsh, P., Lim, K.Y., Long, J.B., and Carver, M.K. (1988) Integrating human factors with system development. In N. Heaton and M. Sinclair (Eds) *Designing End-User Interfaces*. Oxford: Pergamon Infotech.

2

Interfacing the laboratory with the real world

A cognitive approach to colour assignment in visual displays

John Campion

Acknowledgement

The research described in this chapter was carried out under a contract held by the Ergonomics Unit, University College London from the Behavioural Science Division of the Admiralty Research Establishment. The assistance of the Superintendent of this division, Dr. Gardner, and his staff, in the completion of this research, is gratefully acknowledged.

2.1   Aims

2.1.1   Chapter

This chapter describes a cognitive approach to research on the assignment and assessment of colour in computer generated visual displays in naval command systems. It provides an explanation of and justification for the approach, describes the approach and illustrates its use with reference to a research project set up to investigate colour use on submarine tactical plan displays.

2.1.2   Approach

The aim of the approach is to inform system design and development by generating knowledge of colour effects which is both *theoretical* and *objective*. By 'theoretical' is meant that knowledge is based on explanatory principles enabling it to be generalised across instances of potential application; by 'objective' is meant that such principles are subjected to empirical validation.

2.1.3   Research

The research upon which this chapter is based concerns the use of colour in naval tactical plan displays. These displays, together with a set of manual plots, present to a ship's commander a picture of the ocean environment derived mainly from a suite of passive sonars. They constitute the basis for the generation, implementation and evaluation of tactical plans. The display of particular interest here is the Labelled Plan Display (LPD) which provides a synthetic prediction of the disposition of sound sources around the commander's own ship. A schematic display is illustrated in Figure 2.1 to give the reader some idea of the range and type of symbols. A detailed understanding of the display is not required.

Current generation displays are based on cursive technology and are monochrome orange or red. Future generation displays will be based on raster-scan technology able to produce high resolution pictures in a wide range of colours. For example, it is common for computer systems to boast the ability to produce up to 10^6 colours. Clearly, then, the technology now provides no real limit to the number of colours available for use, but the question arises as to which colours should be provided to optimise a system in a given task environment. The question divides into two parts; the first concerns the issue of colour production (how to produce a given set of colours) and is a matter of technology; the second concerns the issue of colour function (for what purpose a given set of colours should be used) and is a matter of psychology. This chapter addresses the second only.

2.2   Approach
2.2.1   Objectivity

We can consider the problem of applying colour knowledge within the Framework provided by Long (Chapter 1, this volume). According to the framework, there is the potential for scientific knowledge to feed forward to new system design and development. However, this potential is seldom realised in the case of psychological sciences and much development takes place through the building of prototyping devices fed by knowledge gleaned from design guidelines, and the intuitive judgements of expert human factors practitioners and users. In such cases the knowledge can be regarded as being implicit rather than explicit and as being distributed among the heads of the experts and instantiated in the form of any prototyping devices built. Such methods may be more or less effective, but they should be regarded as necessary rather than satisfactory, because the implicit and distributed nature of the knowledge renders the system development vulnerable to arbitrary influences such as ill-founded conventional wisdom, ill-founded design standards and pragmatic commercial influences. The wish to mitigate these potential influences is the prime motivation for the objectivity in applied knowledge which is sought in the approach described here. Objectivity is sought in two ways; first by deriving colour assignments from

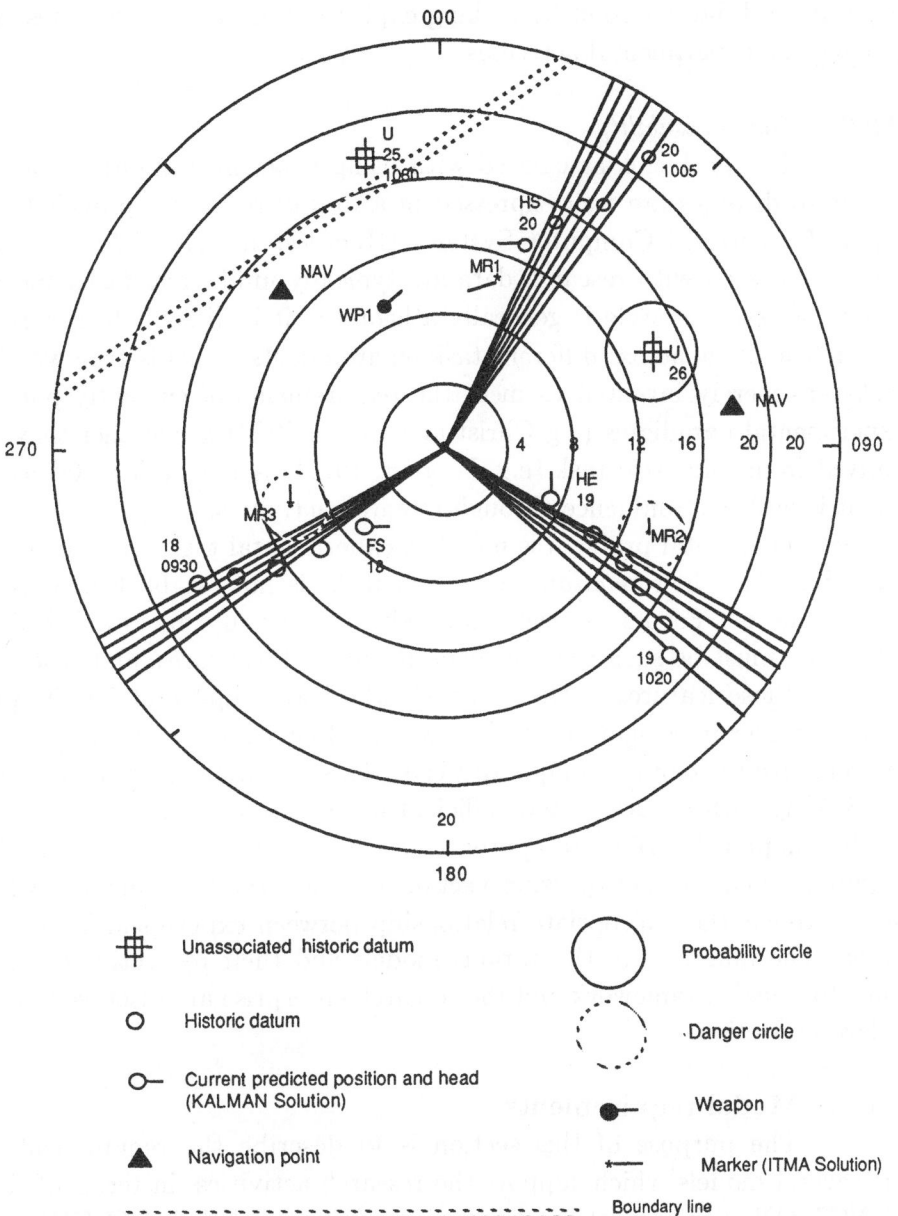

Figure 2.1: Labeled plan display (schematic)

Hypothetical contents of a Labeled Plan Display, upon which the experiment described in this chapter is based, showing the complete range of symbols and typical arrangements and degree of clutter. It is not necessary for the reader to have a full understanding of the meaning of these symbols and the tactical configuration shown is not authentic. Bearing lines (those radiating from the centre of the display) indicate the bearing of sound sources. ITMA and KALMAN are target motion analysis algorithms used to predict various other tactically important parameters of the sources.

experimental data; second by making explicit the analytic structures supporting the experimental activities.

2.2.2   *Generalisability*

The problems associated with using experimental data to assign colour to displays are well expressed in a literature review provided in a report for Ferranti Computer Systems (Huddleston, 1985). In brief, the tasks associated with research data are typically unlike real tasks and because the research style is generally atheoretic, it is difficult to generalise research findings across different task environments. This is true whether tasks are merely invented to meet the requirements of currently popular experimental paradigms (e.g Christ and Corso, 1983) or whether they are derived from some real task (e.g. Wagner, 1977). A real task is defined as one it is wished to inluence through research activities.

The fundamental problem is not that experimental tasks are unlike real tasks but that their *relationship* to real tasks is not established by some explicit means. For an experimental task to be useful, this established relationship must be such that the experiment both captures, and is seen to capture, those features of the real task which are important for the phenomena under investigation; in the case of colour, features of a task which are sensitive to colour coding of displays. This requirement for experimental tasks is both necessary and sufficient if they are to be useful.

This chapter describes an approach which attempts to show how such a requirement can be met by using a set of analytic structures (models) which make explicit the appropriate relationship between experimental and real tasks. The functions of the various models and their respective relationships to Long's Framework and the research enterprise are discussed in the following Section.

2.3   Model requirements

The purpose of this section is to describe the requirements of the several models which support the research activities, in terms of their FUNCTION with respect to ergonomic activities and their STRUCTURE as it relates to the assignment of colour solutions and their assessment in experimental tasks. Both are accomplished using the Framework for Ergonomics described by Long in Chapter 1 of this volume.

2.3.1   *Functions*

The main contribution of the research in terms of the Framework is shown in Figure 2.2. Thus it is concerned, primarily, with establishing an explicit relationship between a real world system and an experimental system (an acquisition representation). This is accomplished by a set of models which are used to describe a real world system and tasks associated with it, extract experimental tasks appropriate for colour investigations, and interpret the data gleaned from them.

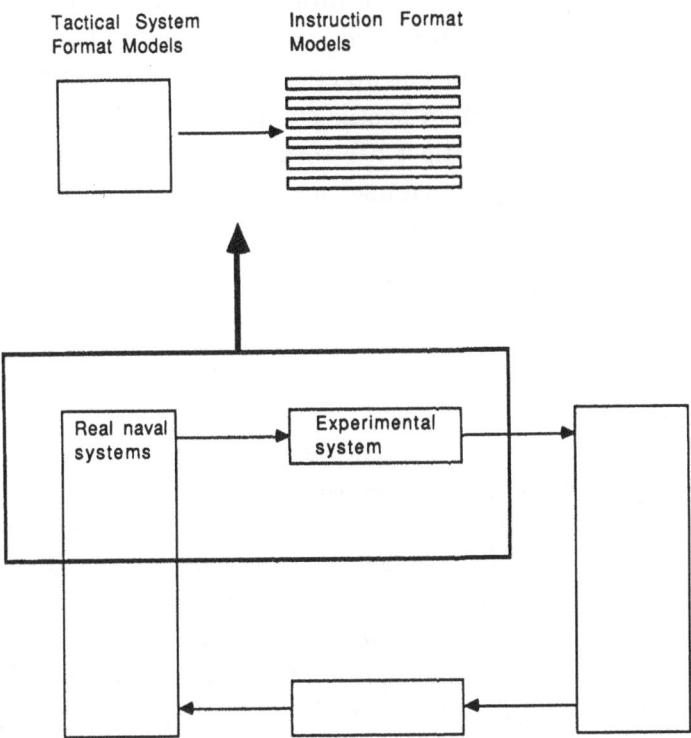

Figure 2.2: Approach and the framework
The approach illustrated in this chapter in terms of Long's framework for ergonomics (Chapter 1, this volume). The SCOPE is shown by the bold-line box and the functions of the two model groups (Tactical System Format and Instruction Format) are shown in the upper part of the figure, where each is shown above the part of the framework to which it relates. Thus, the Tactical System Format describes real systems upon which the research is based, and the Instruction Format is extracted from this and forms the basis of the experimental tasks.

The initial set of models is based on the Tactical System (see following section) and are derived from observation of tasks carried out on a training simulator of the real world system, from interviews with expert operators and from reading manuals. The structure of the models is motivated by the nature of the system itself and also the research requirement, to examine the coding of displays. The second set of models is based on the Instruction Format and is derived from sampling from the Tactical System and re-expressing the sampled part as a number of different levels. The structure of these models is motivated by the requirement to support an experimental paradigm and its associated tasks.

2.3.2   Colour assignment

For the purpose of exposing the requirements of the models for colour assignment, a task can be broken down into units, each one comprising a description of a displayed state (DS) and a description of an operational state (OS) contingent upon it. For example, a DS might be 'best predicted current position of hostile submarine = X' and a contingent OS might be 'preparation for tiger fish torpedo attack'. Another DS might be 'possible zig (i.e. change in course) indicated on Track N' and the contingent OS might be 'carrying out zig drill' (e.g. drop a probability circle on current predicted position). Such task units can be written in general form as contingent rules with the following syntax:

IF (DISPLAYED STATE) THEN (OPERATIONAL STATE)

in which the particular contingencies are established by factors such as the strategic role of the ship, the particular rules of engagement and the status and skill of the person performing the task.

But, such a structure is inadequate for experimental purposes because it is not possible to use it for assigning colour to the display as an independent variable. Thus 'best predicted current position of hostile submarine' is not something which could be *coloured*, because it is not a description of display features, but is a description of a relationship between features of the external world represented on the display. In order to satisfy the colour assignment requirements, the task description would need to contain a DS expressed in terms of display features and decomposed to a level of detail sufficient for colour to be assigned to it.

2.3.3   Colour assessment

The rule structure is also inadequate for assessing the consequences of such a colour assignment. Since a researcher wishing to assess experimentally manipulated colour assignments has to generate a set of contingent rules which are sensitive to colour, and use these to compare alternative colour solutions applied to the decomposed display state, he also requires a higher level description of the task to enable him to compare the consequences for the task of the various colour assignments. The appropriate level of task description will be that at which there are consequences for the task of colour assignment.

In summary, then, the models are required to establish an explicit relationship between real tasks and experimental tasks designed to detect the influences of various colour solutions. In order to achieve this, the models must produce a description of the real tasks at a number of levels. The levels which will be appropriate in this instance will be set by the level necessary for the assignment of colour solutions in the case of the lower level, and by the level sufficient to detect colour inluences in the case of the upper level.

This concludes the account of the requirements of the various models used in the research, the following section describes the models themselves, constructed to meet these requirements.

2.4   Task model − tactical system format

This section describes the first part of the modelling process, that is the part based on the Tactical System. This forms the basis for the sampling and re-expression required for the experiment and described in the subsequent sections.

A General Task Model (GTM), based on a framework derived from a tactical system, was constructed in three phases of modelling, each of which served a different function. Only two are described for the purposes of this chapter. The first phase provides an abstract representation of the system as a set of states and transformations, which sets up the foundation of the GTM. The second phase completes the GTM by mapping onto it the user's knowledge of each of the states and transformations. The exchange of information between these last two components forms the basis for modelling the experimental tasks.

2.4.1   *Tactical system, states expression*

The model (shown in Figure 2.3) comprises three worlds; the external world (e.g. the ocean), the display world of the relevant ship, and the user world. The user receives information about the external world via the display world and also acts on the external world via the display world. On the left of the model, receipt of information about the external world is from the ship sensors, and acting on the external world is accomplished by manoeuvering the ship or firing weapons. The display world comprises sets of media whereby incoming or outgoing information can be made explicit. The same description also applies to the right side of the model, only in this case the source of information is shore-based command, for example an intelligence report. The action would be a report back to shore based command. Left side information is described as *bottom up* and right side information as *top down*.

The states expression of the model (T-System(S)) represents the system as a set of *states* of each of the three worlds connected by *transformations* which describe the change from one state to another. The aim of this abstraction is to describe, not the physical components of the system, but *relationships* between them.

To give an example, the state of the external world 'hostile submarine at twenty thousand yards' might be represented in the display world, on the labelled plan display (LPD), as 'KALMAN symbol for current predicted position near twenty thousand yard range ring'. Note that a single state of the external world could be represented by a number of states of a given display and also by states of a number of displays. Thus, in the example above, the external world state could be represented by a marker on the LPD (see Figure 2.1) and also by bearing rate on another display which shows bearings over time – the time bearing display. It follows that a given state of the external world can be completely described in the display world as the set of all states of displays having a relationship with the state of the external world which are detectable by a user for the purposes of tactical

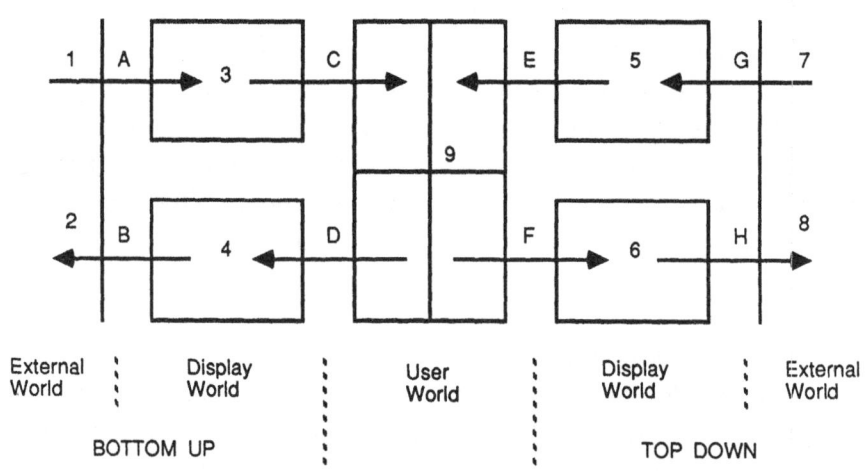

1	Source state	A	Sensor system TF
2	Own ship state	B	Control system TF
3	Display state	C	Visual / Auditory system TF
4	Command state	D	Speech / Manual system TF
5	Top down report state	E	Visual / Auditory system TF
6	Report to top down state	F	Speech / Writing system TF
7	Top down source state (receiver)	G	Radio / Writing system TF
8	Top down source state (sender)	H	Radio / Reading / Administration
9	User states		system TF

Figure 2.3: Tactical system model (S)
The tactical system modelled as a set of states of the system (boxes) and transformations between states (arrows). In the key, numbered items refer to states and lettered items to transformations. TF = transformation due to the system listed.

planning.

The link between external and display worlds is represented as a transformation describing the change from a state of the external world to the corresponding state in the display world. There will be a transformation for each display world state relating to a single external world state. A transformation will be a product of the sensor characteristics interacting with the ocean conditions and with the computational and display coding procedures acting on the sensed data.

The T-System(S) models the user world as a state of a given user. Thus, a component might be 'commander's representation of the hostile submarine marker position'. Note that this is his representation of the display state and not that of the external world state. Unlike the relationship between states of the external world and displays, there will be only a single state of the user with respect to a single state of the display at a given point in time. However, this will not be the case for different points in time, because the particular representation that a user has will depend upon the particular tactical decision subserved by it. For example, if the tactical plan includes avoidance of detection at all costs by a hostile submarine, then the commander's representation of its marker position could be that sufficient to indicate simply the closest possible position, whereas if the tactical plan includes the sinking of the hostile submarine, then the commander's representation could be that necessary to indicate its position within given margins of accuracy.

The link between display and user worlds is represented as a transformation, equivalent to describing the process of perception as a change from one state to the other. The term 'perception' is technically defined as all those processes which serve to transform a given state of the display into a related state of the user, appropriate for subserving a tactical decision. This renders the perceptual transformation very complex, but it has two advantages. First, it exposes parts of the system (such as the physical space between a user and a display) which are important for colour coding; second, it allows a precise definition of perception which is useful for experimental purposes and avoids the problem of having to distinguish, for example, between perception and cognition. Here, the term perception subsumes processes traditionally described as both perception and cognition.

The model, were the contents of the boxes and arrows complete with respect to a tactical decision, would represent the state of the system at a single point in time. More correctly, since the continuous flow of information is actually integrated over time through various mechanisms, including the decision making processes of the user, the model would represent the state of the system integrated over a period necessary for a tactical decision.

2.4.2   Tactical system, knowledge expression

The T-System(S) represents the exchange of information between the user and sources as a set of states and transformations. This can be regarded as describing the *real* state of the system subserving a tactical decision. The user, however, does not have direct access to these states, although he requires knowledge of them in order to carry out his tactical planning. Knowledge of the complete system is acquired via his knowledge of the display states only. The user task can therefore be modelled completely by describing the knowledge he has of each component of the T-System(S). This produces, in the model, an overlay of the T-System(S) (see Figure 2.4). These descriptions, respectively, reflect the system as it is and the system as the user knows it. They provide a potentially complete description of the task of making a tactical decision. The two systems can be regarded as exchanging information via an *executive* which is responsible for controlling this exchange. The executive makes explicit the indirect nature of the relationship between the T-System (S) and the T-System (K); that a user only has knowledge of the external world via knowledge of the display world, for example.

Note that the T-System(K) adheres to the principle of comprising states and transformations, hence boxes and arrows refer to the same classes of components, but they denote states and transformations of knowledge about components and their relationships, rather than denoting the components themselves. The direction of the arrows are reversed relative to the T-System(S) because transformations are directed towards knowledge of the external world. In terms of input and output, these become the current inferred state (CIS) and current projected state (CPS) of the external world respectively. The derivation of each of these via the executive represents the generation of a tactical plan and the implementation of each of them represents the execution of the plan (see Figure 2.4). The state of the T-System(K) constitutes the tactical plan at any given point in time.

The status of the user representations in the two systems needs to be clarified because, although the T-System(K) claims to represent the user's knowledge of the state of the system in contrast to the T-System(S) which represents the real state of the system, the user representation in the latter clearly represents knowledge of the displays in some sense. The best way of distinguishing the two user representations is on the grounds that they represent different levels of knowledge. The higher level can be defined as that sufficient to generate knowledge of a display state and the lower level can be defined as that derivable from a display state and sufficient to generate a higher level state. These relationships will be clarified in the following section.

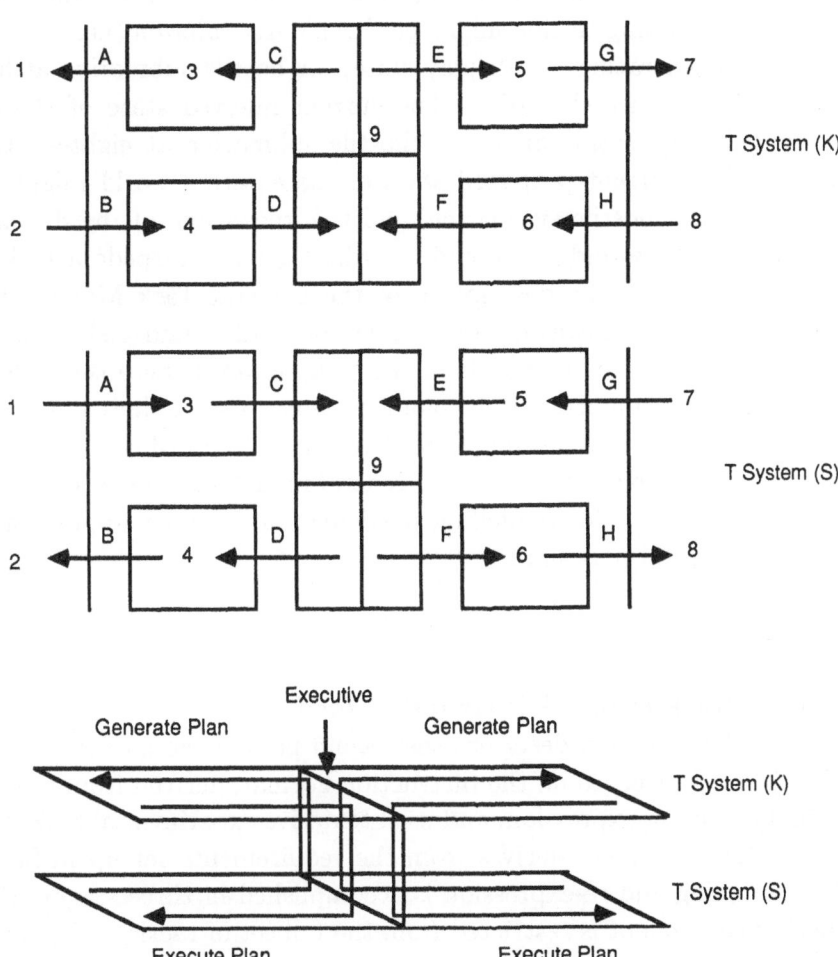

Figure 2.4: General task model

Top: T-System(K)   Centre: T-System(S)   Bottom: General task model showing the relationship between the systems in terms of information flow. The purpose of the executive is to indicate the location and direction information exchange between the systems. Letters and numbers in T-System(K) are equivalent to those in T-System(S) but each term should be prefixed by "Knowledge of —".

The status of the T-System (K) components can be clarified by reference to the hypothetical example used above and the schematic Labelled Plan Display shown in Figure 2.1. Thus, the display state might be 'KALMAN *symbol* for hostile submarine at twenty thousand yards' and the user's knowledge relating to this might be 'KALMAN *solution* (i.e. the computer algorithm generating the display state) for hostile submarine at fifteen to twenty five thousand yards'. The current inferred state of the external world relating to this might be 'hostile submarine at eighteen thousand yards'. The current projected state of the external world might be 'take action to sink hostile submarine' and the knowledge of the display world relating to this would be 'intend to order tiger fish torpedo attack'.

This concludes the description of the General Task Model which can be regarded as a means of structuring command team tasks generally. It does not capture team aspects of the task, however, because these are not thought to be important for colour coding. Also it does not model any particular task, and the function of the model described in the following section will be to sample from the general model and re-express the sample in a form suitable for modelling particular tasks and mapping these onto an experimental paradigm.

2.5   Task model – instruction format

This section describes the second part of the modelling activities, that is the part based on the Instruction Format, derived from the GTM by sampling and re-expression, and used to derive experimental tasks from real tasks. The structure derives from the requirements set up in Section 2.3 and sampling and re-expression is accomplished in three stages. First, the display component is restricted from the full set of tactical displays (which includes the time bearing display etc) to the LPD only and the top down component is converted into a set of instructions comprising an operational state contingent upon some display state; second, the bottom up component of the model which constitutes the display state of the instruction, is expressed as a number of levels of description and each level is specified in terms of a syntax (this re-expressed model forms the Task Model Instruction Format (TMIF), see Figure 2.5); third, the lowest level of description of the TMIF, that of Display Entities, is taken and expressed in terms of a framework which allows tasks and colour solutions to be mapped independently onto a display description in order to form experimental independent variables. This forms the Display Separation Description Framework (see Figure 2.6).

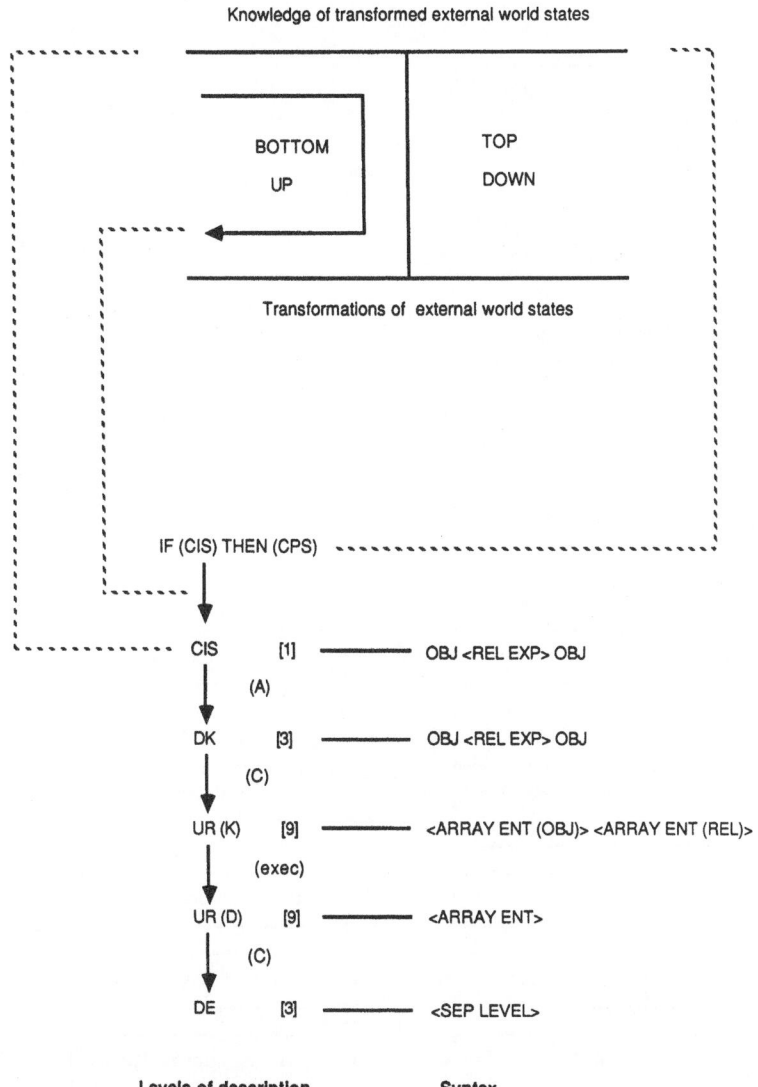

Figure 2.5: Task model instruction format (TIMF)
The relationship between the general task model (GTM), shown schematically in the
upper part of the figure, and the task model instruction format (TMIF), shown in the
lower portion of the figure. Broken lines indicate equivalent parts of the models. Top
down information is converted to a contingent rule and the CIS of this rule is
decomposed into successively lower levels of description. The arrow in the GTM
represents direction of decomposition and hence shows the relationship between the
parts of the GTM and the TMIF. CPS = Current Projected State, CIS = Current
Inferred State, DK = Display Knowledge, UR(K) = User Representation
(Knowledge), UR(D) = User Representation (Display), DE = Display Entities. The
composition of each of the levels is shown in terms of their syntax. Letters and
numbers in brackets show the relationship with T-System(S) and relate to letters and
numbers in figures 2.3 and 2.4

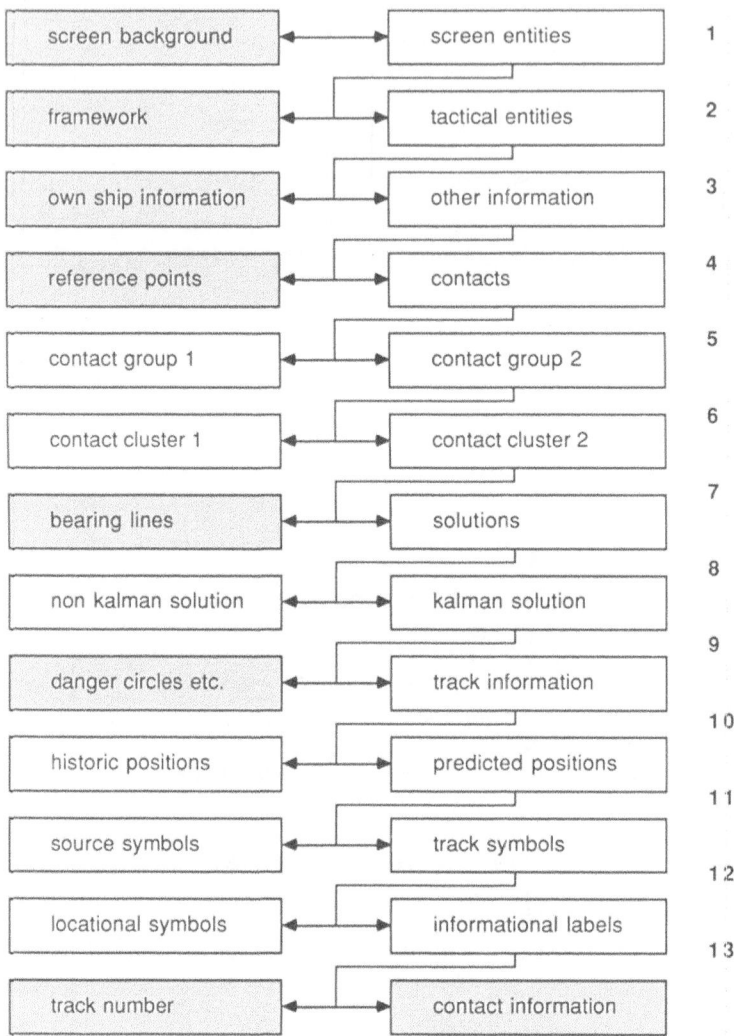

Figure 2.6: Display separation description format

The framework used to derive descriptions for assigning colours and tasks to an experimental display. The terms used in the framework refer to classes of display entity and are hierarchically arranged so that the terms used at one level are superordinate to those used at a lower level. The description is coherent at any level, that is to say that the terms used provide a description that is complete. The terms at a given level are taken to offer potential separations of classes of entity which may be effected by the use of colour. There may be any number of separations at a given level, but only two are shown for the purposes of the framework. At each level, each term is re-described in terms of the lower level terms except for filled boxes which are not re-describable. Only the right hand side is shown as re-described for the sake of clarity.

2.5.1 Format

The accommodation of the right side of the model to non-LPD display information is accomplished by converting all the top down information into a set of instructions of the general form: IF(S) THEN (S') where S is some state of the system and S' is some new state of the system. In the case of the T-System(S), S is some state of the external world and S' some new state of the external world. In the case of the T-System(K), S is the CIS and S' is the CPS. The instruction is not intended to imply that there is necessarily an explicit instruction, but only that there exist states of the system which serve to determine completely an operational state, given some state of the display and the user's task. The instruction format is chosen to represent this for two reasons. First, top down information will frequently be in the form of an explicit instruction (for example 'carry out zig drill'); second, the instruction format lends itself readily for adaptation to an experimental paradigm.

The second part of the re-expression of the model takes the instruction format of the right side and expresses the bottom up states of the left side, which go to make up the CIS, as successively lower levels of description. These are termed, respectively, Display Knowledge (DK), User Representation (Knowledge) (UR(K)), User Representation (Display) (UR(D)), and Display Entities (DE) (see Figure 2.5). The number of levels and the language of description at each level are determined by the function of the model, as explained in Section 2.3. For the purposes of generating experimental tasks from the model, it is assumed that the highest level sufficient to detect any colour influence, is at the level of display knowledge (DK), and that the lowest level necessary to enable colour solutions to be generated is at the level of Display Entities (see Section 2.5.3).

2.5.2 Syntax

Each of these levels has an associated syntax for its expression (see Figure 2.5). The CIS description is an inferred state of the external world sufficient to support a CPS and is written as a proposition establishing a relation or set of relations in the external world. The syntax is:

OBJ ⟨ REL EXP ⟩ OBJ

where OBJ is an object in the external world and REL EXP is a relational expression connecting the two objects. An instance of the example given above could be written as:

HOSTILE SUBMARINE ⟨ IS AT RANGE 10,000 YARDS FROM ⟩ OWN SHIP

Note, that this proposition is a representation of the user's belief about a state of the external world *for the purposes of making a tactical decision.*

Thus, the user's real belief might be that the hostile submarine lies anywhere between 8,000 yards and 20,000 yards range, and the single range 10,000 yards might be an assumption he is prepared to make for the purpose of being able to reason tactically.

The next lower level can be termed Display Knowledge (DK). This has identical syntax, but objects and relational expressions will refer to knowledge about the LPD rather than about the external world. This description is to be regarded as being at a lower level on the grounds that a proposition about the state of the external world can be re-described as a set of propositions about the state of the display world. For example, in the earlier instance, the propositions at this level relating to the single proposition at the higher level, might be:

ITMA SOLUTION (IS AT RANGE 12,000 YARDS FROM) OWN SHIP POSITION
KALMAN SOLUTION (IS AT RANGE 8,000 YARDS FROM)
OWN SHIP POSITION

Note that these propositions describe states of the display as they relate to the external world. So the relational expression refers to the state of the external world indicated by the relations on the display. It does not, for example, have a range expressed in inches. The complete set of propositions at this level is defined as all those necessary to generate a CIS. It should be noted that these propositions are re-descriptions in the sense that they go to make up the higher level description, given the top down information that allows this inference to be made.

The next two lower levels are different in kind from the previous two which both referred to knowledge states. These can be regarded as user representations, but at different levels; UR(K), which is nearer to its respective knowledge state, and UR(D), which is nearer to its respective display state. Both URs relate to sets of display entities. The syntax of UR(K) is ⟨ARRAY ENT (OBJ)⟩ ⟨ARRAY ENT(REL)⟩ where the arrays are two sets of entities. The first is the set of entities comprising the objects, and the second is the set of entities generating the relational expression. Thus, in the example above, the array of object entities would comprise the ITMA marker and the KALMAN symbol, and the array of relational entities would comprise the set of non-object entities in the display used to compute the relation expressed in the UR(K). For example, the ranges indicated by the ITMA marker and KALMAN symbol might be computed by noting the range of the nearest range ring to the symbol and interpolating between the symbol and the next range ring, and either adding or subtracting this amount. In this case, the relational entities would be the respective range rings. In syntactical form, this would appear as:

⟨ITMA MARKER SYMBOL, KALMAN CPP SYMBOL⟩
⟨RANGE RING 1, RANGE RING 2⟩

UR(K) represents all those display entities necessary to generate the relational expression of the UR(K). It should be noted that these expressions refer to the user's representation of the symbols, not to the symbols on the display.

The syntax of UR(D) is similar to that of UR(K) in that it reflects a single array of entities: ⟨ARRAY (ENT)⟩. These are all the display entities necessary to locate a single entity in the UR(K). For example, the user might know that he requires the ITMA marker in order to compute the range of the hostile submarine, but he has first to locate it on the display. His strategy for so doing will depend upon the contents of the display and how familiar he is with it. For example, the user might know that it represents the only hostile contact, and so search for hostile indicators such as labels until he finds one. In this case, the array of entities for the UR(D) might be all symbol labels searched. Alternatively, the user might search the set of marker symbols, seeking the single one that he knows to be hostile, in which case the array of entities would comprise all the marker symbols searched, together with their associated hostility indicators. Note, that there will be a set of entities, and hence an array, for all the UR(K) entities, both object related and those used to establish their relations. Note, again, that these refer to the user's representation of the entities and not to the entities themselves.

The lowest level of description in the model (DE) represents the physical display itself and therefore, in contrast to UR(D), is a description of the entities themselves. In order to relate this description to some colour solution applied to the display the syntax of DE is ⟨SEP LEVEL⟩ (separation level) where the term 'separation' is taken to denote the activity of separating an entity from its background in order for it to be assigned to the UR(D) array. The term 'level' refers to the level at which the separation takes place in terms of a structured description of the display which is presented and explained the following section.

2.5.3   Colour solutions

The function of the lowest level of description is to provide a description of the display entities contained in the higher level descriptions, which will allow the principled mapping of a *colour solution*. A colour solution is a set of rules for applying colour to a display in the design stage of a new system. The syntax of a colour solution rule is:

ASSIGN ⟨COLOUR DESCRIPTION⟩ TO ⟨ENTITY CLASS DESCRIPTION⟩

The analytic structure which allows this assignment is the Display Separation Description Framework (see Figure 2.6). A display is made up of entities such as 'marker' and 'danger circle' (see Figures 2.1 and 2.8) and a colour solution would consist of a set of rules for assigning different colours to different classes of these entities. An entity is precisely defined as the smallest component of a display which could be of tactical importance *and* which could form the basis of a colour solution rule.

The Display Separation Description Framework (DSDF) provides a set of descriptors for classes of display entity, whereas the contents of a particular display, when expressed in these terms, would constitute a Display Separation Description (DSD) (see Figure 2.7). The principles determining the structure of the DSDF are twofold. First, it is structured so that the display entities may be considered in a hierarchical manner from very broad classes at the top to fine details at the bottom. Second, it is structured so that it is complete at a given level of description, that is, it is coherent. Thus, from Figure 2.6 it can be seen that at Level 7 'bearing lines' is coherent with 'solutions' because this level separates what might be called different classes of data source, one being raw sensor data (bearing lines) and the other computed predictions (solutions). These principles determining the structure of the DSDF derive from considering that colour can be said to function at different levels of the display description generated by it. For the purpose of the DSDF, the function of colour is defined as the *separation of classes of entity*. Thus, if hostile entities were coloured red and non-hostile entities green, the function of colour would be to separate hostile from non-hostile entities. The term 'separate' here means only that a property is conferred on two or more classes of entity such that coherence is increased within a class and reduced between classes. The separation into two classes at each level, shown in the DSDF, indicates only that there is a separation, or potential separation, at a given level. In the DSD itself there may be any number of classes separated at a given level. Because the description at a given level is complete a term at one level can be completely described by all lower level terms. Note, though, that not all classes at a given level can be re-described at a lower level. For example, 'bearing lines' at Level 7, which are coherent with 'solutions' cannot be re-described.

The DSDF is used in the following manner. First, for a given display, the DSDF is used to generate a DSD, that is a complete description of the display entities using the DSDF structure. Second, given a set of colours, these are applied to the DSD in a principled fasion. Third, given a real task, this is expressed in terms of the TMIF and applied to the DSD. Thus, the consequences for a task at a given level of the DSD can be determined as a

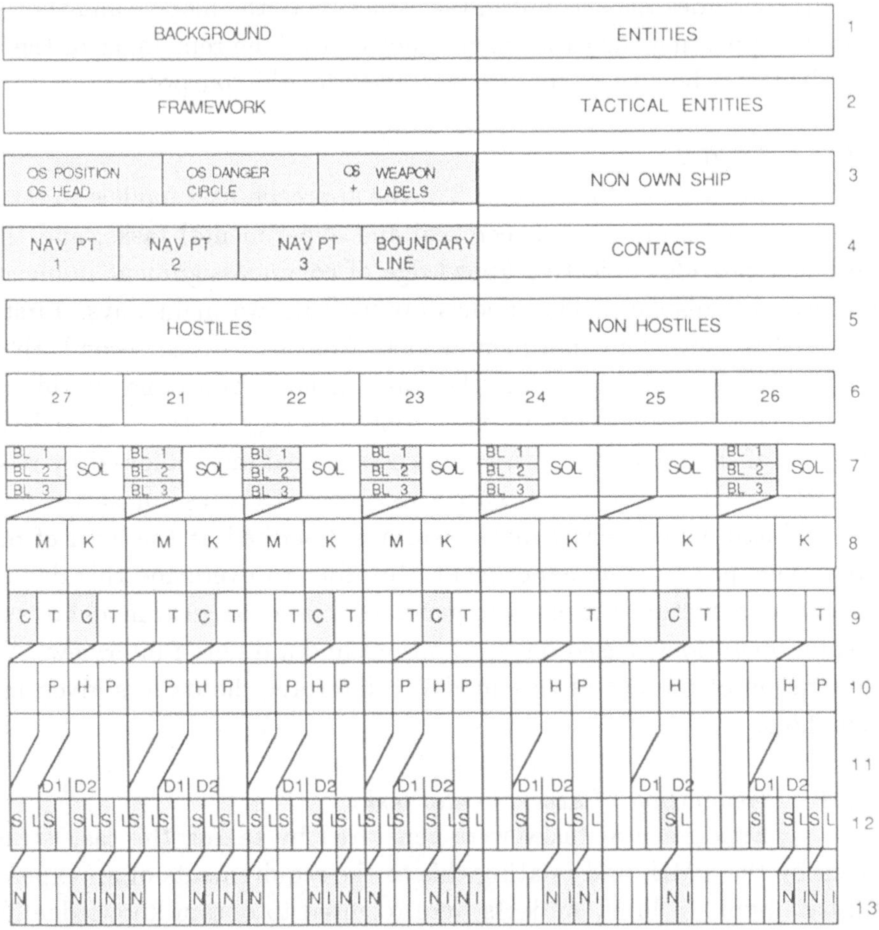

BACKGROUND			ENTITIES				1
FRAMEWORK			TACTICAL ENTITIES				2
OS POSITION OS HEAD	OS DANGER CIRCLE	OS WEAPON + LABELS	NON OWN SHIP				3
NAV PT 1	NAV PT 2	NAV PT 3	BOUNDARY LINE	CONTACTS			4
HOSTILES			NON HOSTILES				5
27	21	22	23	24	25	26	6

BL 1 BL 2 BL 3 SOL · BL 1 BL 2 BL 3 SOL · BL 1 BL 2 BL 3 SOL · BL 1 BL 2 BL 3 SOL · BL 1 BL 2 BL 3 SOL · SOL · BL 1 BL 2 BL 3 SOL — 7

M K · M K · M K · M K · K · K · K — 8

C T C T · T C T · T C T · T C T · T · C T · T — 9

P H P · P H P · P H P · P H P · H P · H · H P — 10

— 11

D1 D2 · D1 D2 · D1 D2 · D1 D2 · D1 D2 · D1 D2 · D1 D2 — (11)

S LS S LS LS LS · S LS LS LS · S LS LS LS · S LS L · S S LS L · S L · S S LS L — 12

N N N N · N N N · N N N · N N I · N N I · N I · N N I — 13

Figure 2.7: Display Separation Description for experimental display
The complete set of entities contained in the experimental display (see Figure 2.8) expressed in terms of the DSDF (see Figure 2.6). A display entity is represented by a filled box. BL = bearing line, SOL = solution, M = marker, K = Kalman, T = track, C = circle, P = predicted position, H = historic position, D1 + D2 = datums, S = symbols, L = labels, N = track number, I = contact information, NAV PT = navigation point, OS = own ship.

function of the level of the colour solution. Conversely, for a given colour solution level, its consequences can be determined as a function of the level of the task.

This concludes the account of the structure of the models and their function in supporting the experimental activities. The remainder of the chapter illustrates how these models were used for this purpose.

2.6   Model use

As described in Section 2.3, the approach uses models to establish an explicit relationship between real and experimental tasks, and the experiment provides objective knowledge of colour assignment influences in tactical plan displays. The models are used in two main ways. First they are used to generate experimental tasks from real tasks; second, they are used to interpret the data from the experiment. A third and subsidiary use was to construct a simple but explicit methodology for the system designer. Because of the relatively restricted access to real tasks allowed by research logistics, and because the research necessitated the use of naive subjects, the real and experimental tasks were not described at the level of mental operations prior to the experiment. In order to overcome this deficiency, the model structure was used to generate a psychological model of subject mental operations, based on observations and a de-brief interview. A brief illustration of the use of this model to interpret the data is given in Section 2.6.3.

2.6.1   Generation of experimental tasks

Experimental tasks were generated by taking the task structure given by the TMIF and relating this to the requirements of an experimental paradigm. In brief, subjects were required to obtain knowledge from a simulated tactical plan display under time pressure. The independent variable was colour solution and the dependent variable was response time. In terms of the TMIF, the dependent variable was expressed at the level of display knowledge (DK) in order to meet the criterion set up in Section 2.5.1; that is the highest level at which colour has an influence. DK consisted of two types; the configural relationship between two entities, expressed as some quantity (for example, the distance between the own ship and a given track CPP); and the identity of an entity satisfying the criteria given by some quantified configural relationship (for example, the track number of the closest contact). These two classes of DK arise from the syntax of DK in propositional form. Such an expression, in its three terms, completely defines a state of the world so that, given a true state of the world, the presentation of two terms defines the third. Thus, if the true state of the world can be represented as:

OBJ(1) ⟨REL EXP⟩ OBJ(2)

an interrogative can be written as:

:QUERY: ⟨REL EXP⟩ :GIVEN: OBJ(1) . OBJ(2)

or as

:QUERY: ⟨OBJ(1)⟩ :GIVEN: ⟨REL EXP⟩ . ⟨OBJ(2)⟩

An example of the former would be:

WHAT IS: ⟨THE RELATIONSHIP WITH RESPECT TO DISTANCE⟩
:GIVEN: ⟨HOSTILE SUB⟩ . ⟨OWN SHIP⟩

An example of the latter would be:

WHAT IS: ⟨THE OBJECT⟩ :GIVEN:
⟨THE RELATIONSHIP WITH RESPECT TO DISTANCE⟩ . ⟨OWN SHIP⟩

A set of 40 propositions was generated which captured simple relationships shown on the experimental LPD. These particular propositions were generated from knowledge gleaned from observation of the trainer task, from interviews, and from informed intuitions. These were converted into interrogative format using the two classes of transformation described. Three modes of operation were defined by timing in relation to the display; pre-display operations, display operations, and post-display operations. The impact of colour on each of these was determined by the incorporation of each mode in the experimental paradigm by dividing each trial into three correspondingly timed sections.

A characteristic of LPDs, which makes them different from displays typically used in laboratory tasks, is their different sources of redundancy. The two major sources are the configural constraints placed on display contents due to tactical and strategic level influences, and the learning of the display by a command team member due to the slow evolution of scenarios. Both of these sources were catered for by having a single display which was presented repeatedly over the 40 trials. Thus, the 40 questions asked of each subject all concerned the same display, and the intention was that the subject's increasing knowledge of the display should be used to derive the information required by a question, and that the effect of this on colour coding should be recorded in terms of the changes in performance over trials. A consequence of this strategy, however, was that the generalisability of the findings is restricted with respect to the range of tactical scenarios.

The display contents were derived from observations during a scenario on the training simulator, but the degree of clutter was increased by adding additional contact clusters. The Display Separation Description relating

to the experimental display is shown in Figure 2.7, and the display itself in Figure 2.8. Three colour solutions were derived for the display,
a monochrome (red) version and two coloured versions. The two coloured
versions used the three colours, red green and blue available on the BBC
microcomputer software and the display contents were constructed using
the BBC Computer Aided Design package, BITSTIK.

The effect of colour at two solution levels was investigated; Level 5
and Level 7 of the DSD (see Figures 2.6, 2.7 and 2.8) , together with a
monochrome (red) control condition. Sixteen subjects were tested, and each
subject performed under one colour condition and a control monochrome
condition. The usual measures were taken to mitigate the confounding effects of learning and colour condition.

2.6.2   Subject mental operations

To understand fully the function of colour it was found necessary
to construct a model of the subjects' mental operations employed in the
experimental task (the subject model). This was derived from observation, spontaneous comments and structured debrief interviews. Figures 2.9
and 2.10 show how the task model and subject model were used to relate the real and experimental tasks. The task model provides a common
descriptive framework for both real and experimental tasks, and the subject model expresses the redescription rules linking the different levels into
mental operations intended to have psychological reality.

The syntax at the DK level is OBJ ⟨REL EXP⟩ OBJ, and forms the structure of the answers to the experimental questions (see Figure 2.10). This is a
functional level of description for the objects (for example, KALMAN SO-
LUTION) and not the display symbol associated with it. The latter would
be contained in the next level down, the UR(K) level. The syntax of the
mental operations responsible for the transformations between these levels
is ⟨feature list⟩ ⟨computational algorithm⟩ ⟨metric⟩, where the feature list
comprises the list of features used to fix the objects for the purpose of computation, the algorithm is the process by which the relationships between
the objects is computed, and the metric is the means by which this is quantified and made explicit (see Figure 2.9). Note that the term "feature" is
used here to denote that part of a display entity, or set of entities used to
define the display object for the purposes of computation. It should be distinguished logically from a display entity, although it could, in fact, be a
display entity.

For the relationship between object features to be computed, the features
have first to be located on the display. This is achieved by searching the
display for a target feature. A target feature may or may not be identical

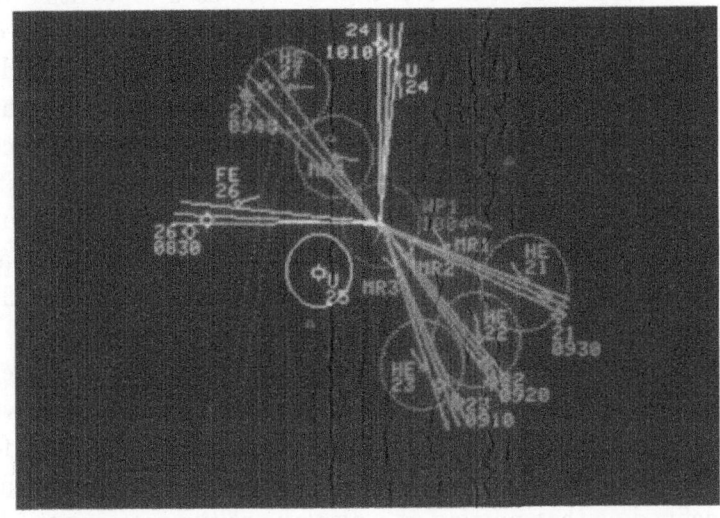

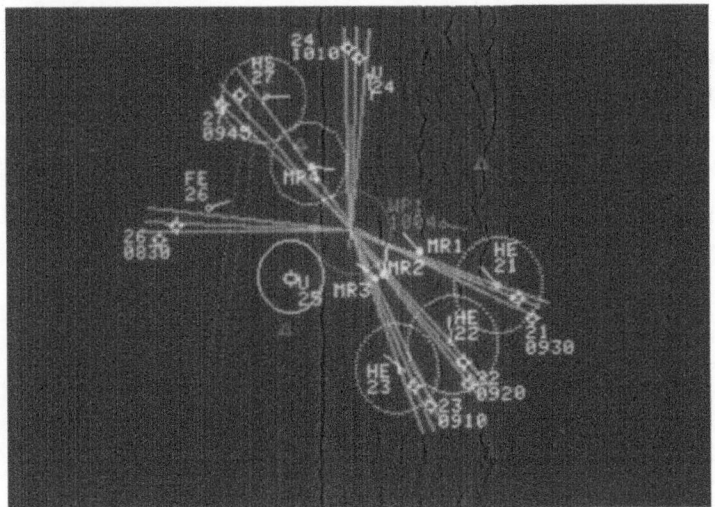

Figure 2.8: Experimental displays
The single display used in the experiment showing the two colour solutions (Level 5 –
top, Level 7 – bottom). The display was generated on a BBC microcomputer. Note
that the colours shown here are not true representations. The monocrome version
used as a control was red only.

This figure is available in colour for download from www.cambrige.org/9780521204842

to an object feature. It is defined as that feature used to identify an object for the purposes of searching the display in order to locate it. In order to locate such a target feature, a set of features have to be searched for the target. Thus the syntax at this level is ⟨target feature list⟩ ⟨search feature list⟩ ⟨locating algorithm⟩. Note that the search feature list will not comprise all the features of the display because of the top down information the subject may be able to employ. For example, knowledge accumulated of the display might enable the subject to restrict his search to a particular area, and within this area, the search feature list would comprise the set of features known to be sufficient to identify the target.

Both target and search features have to be discriminated or separated from the background against which they are set. This forms the lowest level of description, and has the syntax ⟨separated feature⟩ ⟨background feature⟩, each expressed as a level of the DSD (see Figure 2.10). A discriminated feature is defined as that portion of a display entity necessary to identify a search or target feature, and a background feature is defined as that portion of a display entity or entities having an influence on the discrimination. The syntax at this level is thus: ⟨DSD LEVEL(f)⟩ ⟨DSD LEVEL(b)⟩. A complete example of the mental operations at the different levels is given in Figure 2.10.

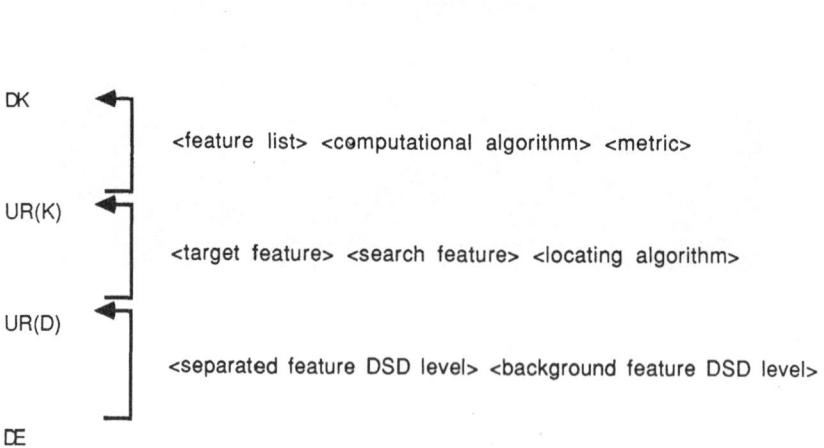

TMIF   **Mental Operations**

DK

              <feature list> <computational algorithm> <metric>

UR(K)

              <target feature> <search feature> <locating algorithm>

UR(D)

              <separated feature DSD level> <background feature DSD level>

DE

Figure 2.9: Subject model
Model of subject mental operations involved in the experimental task of deriving display knowledge from the display. On the left are shown the different levels of Task Model Instruction Format (TMIF), and the mental operations are those necessary for making the transformations between the levels. DK = Display Knowledge, UR(K) = User Representation (Knowledge), UR(D) = User Representation (Display), DE = Display Entities.

	TMIF	SYNTAX	EXAMPLE
Task Model	CIS	OBJ <REL EXP> OBJ	HOST. VES. <10,000 YDS> OWN S.
	DK	OBJ <REL EXP> OBJ + OBJ <REL EXP> OBJ	ITMA <12,000 YDS> OWN S. + KALMAN <8,000 YDS> OWN S.
	UR(K)	<ARRAY ENT (OBJ)> <ARRAY ENT(REL)>	<MR1+OS CPP> <R.RING1+R.RING2>
	UR(D)	<ARRAY ENT>	<MR3+MR2+MR1>
	DE	<SEP LEVEL>	<13+13+13>
Subject Model	DK UR(K)	<f1+f2+ > <ALG> <METRIC>	<CPP(HE23) + CPP(OS) + RR1 + RR2> <COUNT RANGE RINGS> <MULTIPLY BY 1,000 YDS>
	UR(K) UR(D)	<tf1+tf2><sf1+sf2><ALG>	<HE23><HE -><SCAN ANTI C'WISE>
	UR(D) DE	<DSD LEVEL(f)><DSD LEVEL(b)>	<LEVEL 13><LEVEL 7>

Figure 2.10: Task model and subject model – syntactical relations
Task Model Instruction Format (TMIF, upper part of figure) and subject model
(lower part of figure). The syntax of both TMIF and subject model are shown on the
left, and examples of each on the right. OBJ = Object, REL EXP = Relational
Expression, DSD = Display Separation Description, f = Feature, sf = Search Feature,
tf = Target Feature, HOST. VES. = Hostile Vessel, OWN S. = Own Ship, ITMA =
Improved Target Motion Analysis, MR = Marker, OS = Own Ship, CPP = Current
Predicted Position, R. Ring = Range Ring, HE = Hostile Escort, ANTI C'WISE =
Anti Clockwise.

2.6.3   Interpretation of data

There were four groups of subjects based on the two colour solutions (Level 5 and Level 7) and the two trial block orders (monochrome first and colour first). The mean response times (RTs) for each of these groups are plotted in Figures 2.11 and 2.12, together with their associated standard deviations (SDs). RTs are plotted for time periods T1, T2 and T3 (pre display, during display and post display, respectively). It was thought likely that learning and/or colour would have the effect of reducing the variance of response times across trials, so the SD across trials, for each subject, was computed and the mean for each group plotted along with the RTs. There were thus two types of SD plotted; the SDs of the means across the four subjects, given that each subject had a single mean computed across trials (white bars), and the mean of the SDs across the four subjects, given that each subject had a single SD computed across trials (black bars).

A colour effect would show in the data as a faster RT in the second trial block when it is coloured than when it is monochrome. By this criterion, the data suggest that the colour effect in this paradigm is restricted to Level 7 and time period T2 (that is,the period when the display is present). In both Level 5 and Level 7 solutions own ship entities are blue, but in the Level 5 solution hostile entities are red and friendly entities are green, whereas in the Level 7 solution all bearing lines are red and all other entities, irrespective of hostility, are green. This suggests that the effect of colour is to aid the lowest level operations concerned with discriminating features from the background. An example might be discriminating the CPP of Track HE23 against the background of bearing lines. This is supported by the subjective reports of subjects in this group that colour helped "pick out the minute aspects" (S 14). The complete set of evidence suporting this conclusion, is based on statistical inference, individual trial data, and observation of subjects behaviour, but this is not presented here.

In summary, given a display in which contact clusters are well separated and most of the clutter resides within the clusters, colour seems to help in making clear important distinctions between features which, in monochrome, remain obscured. This could be termed a *decluttering* effect, and would be contrasted with a potential *grouping* effect which would imply the making coherent a set of widely spaced features necessary for some judgement. Such an effect would be expected to show up as an improvement in performance with a colour solution applied at Level 5. The evidence here suggests that, instead, colour at this level is a hindrance rather than a help. This would accord with the above interpretation because a Level 5 solution implies that fine grain discriminations have to be made against same – colour backgrounds. Furthermore, if a number of these

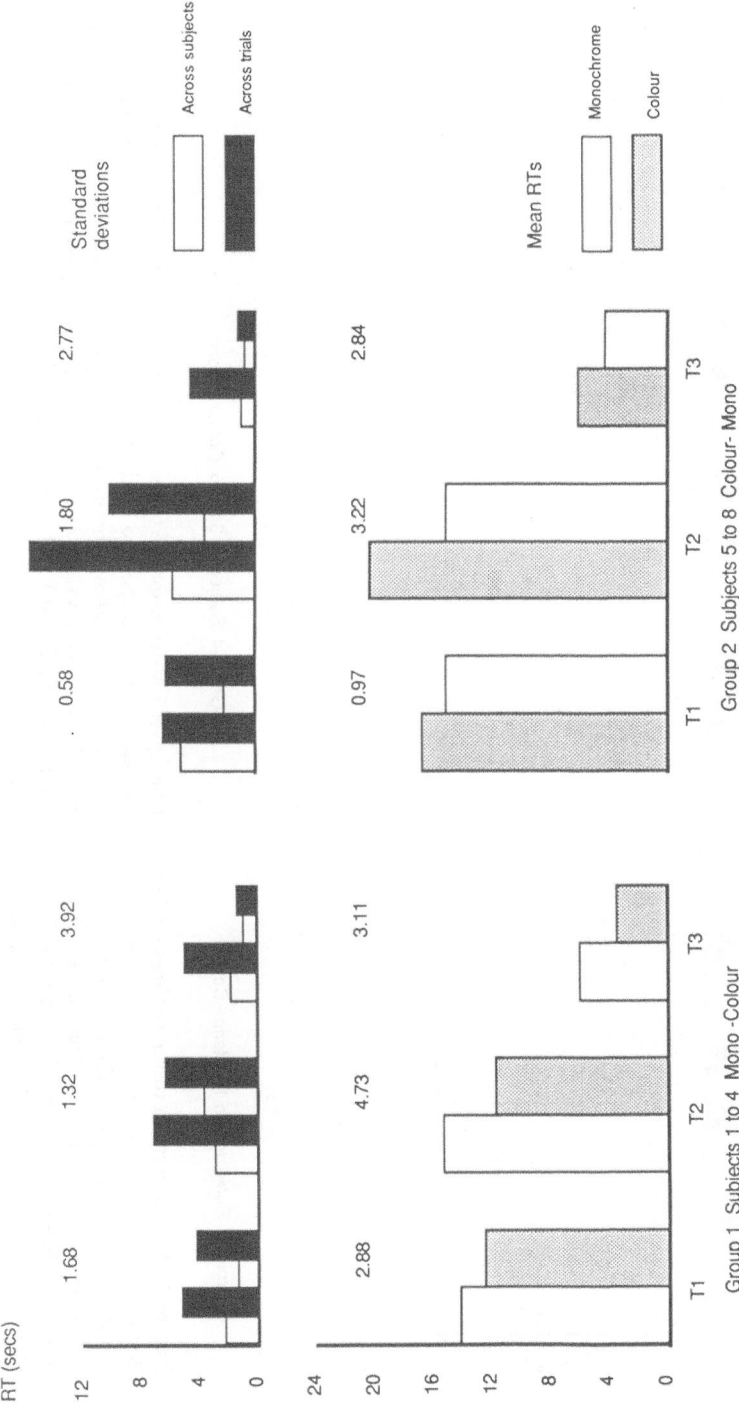

Figure 2.11: Group mean response times - colour solution Level 5 Mean response times (RTs) in lower part of figure and standard deviations (SDs) in upper part of figure. Each pair of bars represents the two trial blocks, and the positioning in the figure represents the sequence in the experiment. The white and black bars above the means represent the two standard deviations computed for the respective trial blocks. Numbers above bars indicate values of t statistic for the difference between the pair of means (black bars for SDs). Critical values for $p < 0.05$ are 3.25 (1 tail) and 3.18 (2 tail).

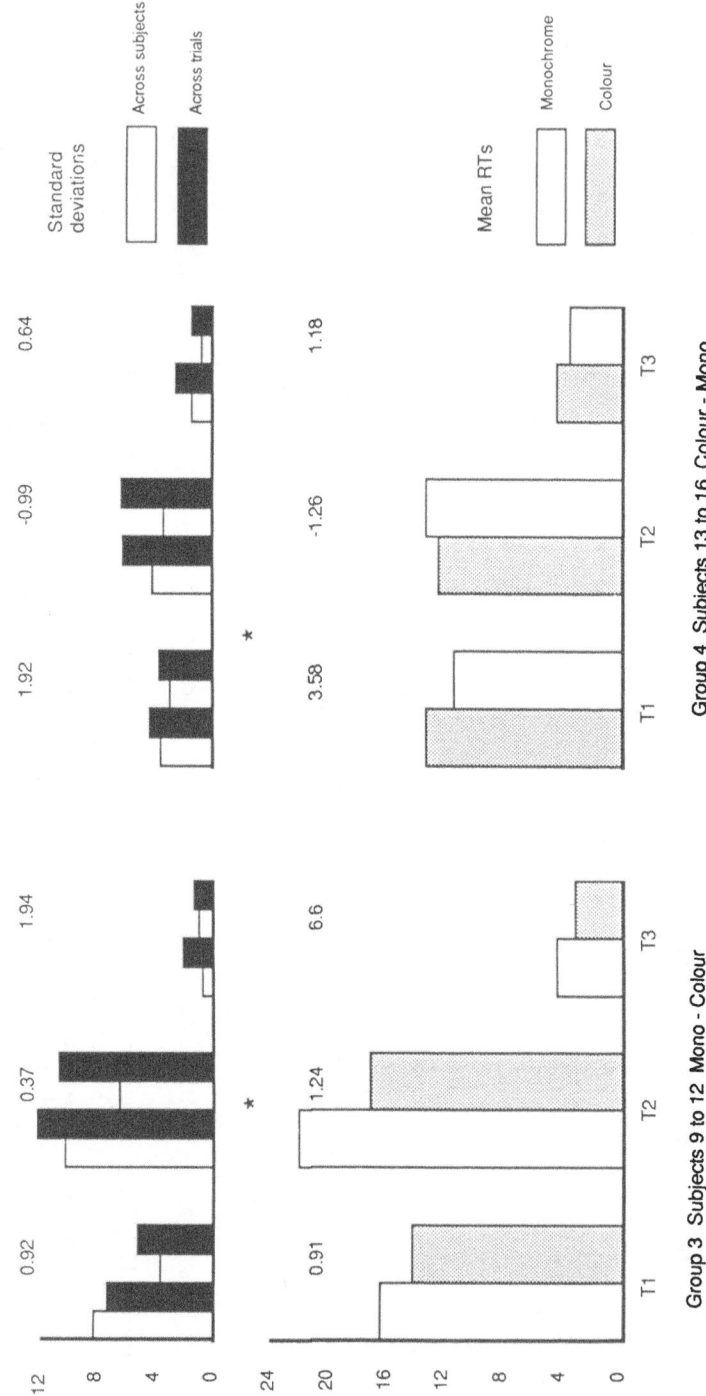

Figure 2.12: Group mean response times - colour solution Level 7 Mean response times (RTs) in upper part of figure and standard deviations (SDs) in lower part of figure. Each pair of bars represents the two trial blocks, and the positioning in the figure represents the sequence in the experiment. The white and black bars above the means represent the two standard deviations computed for the respective trial blocks. Numbers above bars indicate values of *t* statistic for the difference between the pair of means (black bars for SDs). Critical values for *p* < 0.05 are 3.25 (1 tail) and 3.18 (2 tail). Asterisks mark the necessary and unique difference in sets of means indicating a colour effect.

have to be made for a given judgement, the different monochrome environments provided locally for each discrimination could interfere by requiring different discriminating strategies to be used in each case. This, however, is conjecture and requires further exploration and validation.

2.7   Summary and conclusions

2.7.1   Research

The research derived experimental tasks from real tasks using a set of models. The data from the experiment are described and a model of the subjects' mental operations constructed to aid their interpretation. The data suggest that colour aids performance, but this is restricted to the period when the display is present (T2) and the solution applied at Level 7. The model suggests that this is due to fine grain separation of task related features from a cluttered background, rather than to any higher level grouping effect.

This rudimentary analysis of the data was used to illustrate the sort of output produced by the research rather than to provide substantive recommendations for colour use in displays. However, the data provide some pointers towards changes in current guidelines and indications for the direction of future research. With regard to current guidelines, they suggest, for example, that the frequently made, yet unsupported, assertion that hostile contacts should be coloured red as an overriding priority may need qualification and that the consequences of such a policy for different aspects of the task need to be assessed in order to be set against any possible benefits the "hostiles red" solution may produce. Thus, if a colour solution is implemented to produce homogeneous red within contact clusters and homogeneous other colours within other clusters, discrimination of fine grain detail is likely to be impaired and overall performance might be degraded. Whether this is in fact likely to be the case needs to be checked against the requirements of real tasks at the different levels identified in the model, and also the degree of local clutter and resolution of display entities produced by the technology used.

2.7.2   Approach

The intended product of the research, given the approach described in Section 2.2, was to generate applicable colour knowledge which was objective and generalisable. The product fell short of this ideal, so it was recommended to the navy that it be used, either to assess the approach with a view to progressing it through further research, or that the approach itself be used as a simple but explicit methodology for aiding prototyping. As such, it could be used either 'online', to support an experimental paradigm, as in this study, or 'offline', as an aid to discussion. The methodology comprised a set of procedures, a summary of which are provided here:

1. Consider the main display likely to benefit from colour coding.
2. Construct a prototypical display scenario, together with ones that are respectively more and less cluttered.
3. Select a palette of colours suitably constrained.
4. Construct three sets of colours from the palette.
5. Construct DSDs for each of the three displays in (2).
6. Construct a colour solution rule (CSR), for applying the colour sets to the DSD.
7. Apply the the CSR to each of the DSDs in (5).
8. Construct each of the display scenarios according to their respective colour solutions.
9. For each scenario, construct a set of tasks at the instruction level (as defined by the task model).
10. Construct a suitable paradigm if the methodology is to be used 'online', or a set of 'thought experiments' for structuring discussion, if the methodology is to be used 'offline'.
11. Apply the various candidate colour solutions and consider their effects on the tasks.
12. Perform iterations through the foregoing as necessary to arrive at a satisfactory design solution.

2.7.3   Chapter

This chapter has identified the desirability of achieving objectivity of knowledge to be applied in system development, where objectivity is defined as being subject to public validation. A means of rendering knowledge objective is to have it based on experimentation, but in order for experimental data to be applicable it is necesary for it to be generalisable across task environments. In order to achieve this end, it is suggested that experimental tasks need to have a relationship with real tasks such that they capture, and are seen to capture, the features important to the investigation. One way of achieving this is to establish such a relationship be means of a task model.

The use of such models in a particular research domain is described in which colour assignments to naval tactical plan displays are assessed. The models are used to derive experimental tasks from real tasks and interpret the data. It is described how the approach could be developed into a methodology for, aiding prototyping or progressing to deliver substantive recommendations for colour use which would meet the criteria of objectivity and generalisability.

References

Christ, R.E. and Corso, G.M. (1983) The effects of extended practice on the evaluation of visual display codes. *Human Factors*, **25**, 71-84.

Huddleston, J.H.F. (1985) Colour coding for computer driven displays. Report 5370, Issue 3. Ferranti Computer Systems, Bracknell.

Wagner, D.W. (1977) Experiments with color coding on television. USN Report NWC TP 5952; China Lake, Calif.

3

Constructing appropriate models of computer users: the case of engineering designers

Andy Whitefield

3.1 Introduction

An interactive computer system can be thought of as comprising two components: a user and a computer program. If the system is to perform its task well these two components must be suitably matched. The system developer has different options for improving the suitability of each component for the other: users can be selected and trained to suit the program; programs can be structured to suit their users. In structuring programs, developers must understand the users sufficiently well to decide whether the proposed structure will be suitable. This understanding could be gained in a number of ways. One approach, currently under investigation by a number of researchers, is that appropriate descriptions or models of users will help developers decide what is a suitable match. This of course begs the question of what is an appropriate model of a user. This chapter addresses the question of how to construct an appropriate model of a user for a given task. This is done by means of a case study involving the construction of a model of engineering designers. After briefly considering the relevant aspects of design and of computer aided design, the main part of the chapter discusses the derivation, development and content of a model of the engineering design activity. The final two sections describe the model's uses and an attempt to assess it.

To ensure clarity, the term 'designers' will be used to refer to engineering designers (the users in computer aided design systems) while the term 'developers' will be used to refer to those who write and produce systems (and would therefore be those interested in a model of the users).

3.2 Background to design and to computer aided design

Before discussing the model, it is necessary to provide some background to what is known about design and to introduce the technology of computer aided design (CAD). The design background feeds into the development of the model. The CAD background is to give the reader some appreciation of the capabilities of CAD programs, since it is with these that the designer interacts and also since it is CAD system development that is the model's intended application.

3.2.1 Design

Design is a poorly researched activity. It is only really in the last 25 years that any substantial efforts have been made to study and understand it. Prior to that the design community saw it as an intuitive and holistic skill resistant to analysis, and the psychological community, partly restricted by the behaviourist approach, had neither the theoretical nor the methodological tools to cope with it. The sorts of design research that have been conducted in the last 25 years are well represented in Cross (1984). Lera (1983) reviews specifically studies of "the design process and designer behaviour".

The part of this research that is relevant here is the evidence on individual design activity; this is what any model of design should account for. As it happens, this evidence is very sparse. The following criteria were used to select from the literature the relevant findings of empirical studies of design:

- the activity should come under the following definition of design: *design is the creation of specifications to construct objects that satisfy particular requirements*. This definition is based on that of Stefik, Aikins, Balzer, Benoit, Birnbaum, Hayes-Roth and Sacerdoti (1982).
- all or most data should be collected at the time of designing
- the methodology should be clearly reported and replicable

The purpose of these criteria is to restrict consideration to studies that are relevant and likely to be reliable. Even these modest criteria reduce the acceptable studies to an extremely small number. If one includes a number of studies that fail to meet one criteria or another, the evidence can be summarised as follows.

There is convincing evidence that designers do not proceed by formally analysing the problem and logically deriving a solution from this, in the way suggested by the early design methods writers. Design problems are too complex and too poorly specified for that. Design is more of a dialectic between the generation of possible solutions and the discovery of the constraints operating on the solution space (Henrion, 1978; Akin, 1978). The notion of designers as solution-focused, in that their prime activity is to suggest and test potential solutions, is a useful simple characterisation (Lawson, 1979). One way in which designers discover constraints is by the use of a variety of representations. Several are typically used during a design, and each is appropriate for different kinds of constraint (Eastman, 1970; Akin, 1978).

The size and complexity of design problems have at least three consequences. Firstly, the problems cannot be comprehensively specified. Secondly, they will not have a single optimum solution, but rather a range of acceptable solutions, each with a different trade-off between its advantages and disadvantages. Thirdly, the problems cannot be treated as a whole, and must be decomposed into subproblems. The manner in which this breakdown is accomplished is a very important component of skill in design. It might best be thought of as the strategic, domain-independent part of the skill. More skilled designers decompose the problem into more levels, and are more likely to consider the subproblems in a breadth-first rather than a depth-first order. They thus work up an increasingly detailed solution fairly evenly through different levels by solving subproblems in turn (Jeffries, Turner, Polson and Atwood, 1981; Adelson, Littman, Ehrlich, Black and Soloway, 1985).

There is some suggestion that, at least for software design, individual differences can be considerable (Malhotra, Thomas, Carroll and Miller, 1980). And finally, there is evidence that the practicality and originality of designs are negatively correlated, and that at least their practicality can be improved by stimulating access to relevant knowledge (Malhotra *et al*, 1980).

This body of evidence presents a problem to the computer system developer attempting to understand designers. The research is not extensive and has employed a variety of methodologies and investigated several different activities. There is no accepted single coherent picture, and certainly not one that presents this information in an appropriate form for the computer system developer to utilise. Any model of design the developer might use should at least be consistent with this evidence.

3.2.2   CAD

It is important to realise that CAD is only one part of a much larger move towards what is usually called CAE (computer aided engineering). This covers computer applications in production and management as well as in design. Although CAD on its own can have economic benefits, principally in reduced product lead times and increased designer productivity, it is in the context of CAE that its biggest potential advantages lie. This wider context allows one to realise that CAD did not develop from, nor is it implemented for, a simple desire to ease the designer's task. It is the product of a complex set of interacting needs and technologies, extending beyond the immediate design task, and there is no *a priori* reason to believe, simply because of its name, that CAD actually helps designers.

The term CAD is most often applied to the use of interactive graphic computer systems during the design process. This is the meaning that will be used here. Other non-interactive or non-graphic computer design aids are available, particularly for lengthy calculations and occasionally as a designer substitute.

In its simplest form, a CAD system involves the use of a graphic display as a kind of electronic drawing board. In the same way that a word processor allows an author to create, manipulate and store text, CAD allows the creation, manipulation and storage of standard technical drawings in orthographic projections. The designer's principal input is with some form of pointing device, and completed drawings to various scales can be output on a plotter. To enter small details, the user can 'navigate' the drawing by expanding parts of it to fill the whole screen.

At a more complex level, most systems have facilities for integration into manufacture and production control, and these are important determinants of their effectiveness. Additional design functions that may be included are features like simulating performance, checking the product against design principles, and automatic dimensioning of drawings. Most systems allow the development and display of three-dimensional 'wire frame' views of the product.

The most sophisticated current systems go further still. Using what are known as solid modelling techniques, they build up a complex and comprehensive description of the product. As well as describing the three-dimensional shape, this model can include information about material properties, surface textures, cost, and so on. The model can be displayed in perspective views from any angle, with colour and shading information included. In addition, it can be interrogated to supply information such as the product's centre of gravity, weight and volume, or to assess the accuracy of the fit between different components.

The claimed benefits of CAD for the designer (as opposed to the company) vary according to who is making them, but the following are all common:

- they perform complex calculations quicker and with less error
- they considerably reduce repetitive drawing time
- they allow design changes to be implemented more rapidly
- they allow more alternatives to be considered
- they reduce design errors
- they increase designer productivity

Clearly the capabilities of current systems are diverse and they present the designer with more possibilities than a conventional drawing board. But few design tasks have yet been automated and this leads to a high level of user-program interaction. The program interfaces consequently tend to be very complicated. Increasing computer power and developments in artificial intelligence are likely, in the short term, to increase rather than to decrease this complexity. The CAD system developer's problem is to select and structure a set of program functions that will be of maximum help to the designer. Understanding the designer's task so that the program choices can be justified would be an important part of solving that problem.

3.3   Derivation of the model

3.3.1   *Selection of Hearsay-II*

The problem addressed here is to select an appropriate form for a model of the engineering design activity. There are two potentially important constraining influences on that form: the task it is intended to describe (i.e. the psychological evidence on design just discussed) and the task for which it is to be used. The former influence will be discussed throughout this section. Regarding the latter, the model is intended to serve two purposes, neither of which turns out to be a strong constraint on its form. Firstly, to compare the engineering design activity with and without computer aids (see Whitefield, 1986a); the main requirement here is that at least two related but discriminable forms of the model must be available. And secondly, to improve the organisation of CAD systems during their development. This is not a strong constraint because very little is known about how system developers might use human-computer interaction models (see Whitefield, 1986b for a discussion). Some researchers have made assumptions about this. Thus Card, Moran and Newell (1983) assume that what developers need are approximate calculational models, and they go on to present the Keystroke Level Model as an example. In the face of uncertainty about what developers want and could use, it was decided to let the form of the model be influenced most clearly by the task it is intended to describe. The main implication of its use by developers is probably that the model should not be too complicated.

Note that for these purposes, there is no requirement that the model be psychologically real, in the sense of being an accurate representation of the actual mental structures and processes of the designer. Rather, it is intended to be what one might call psychologically appropriate. That is, it should characterise the mental activity of design for its own purposes, in a way that is as far as possible consistent with the psychological evidence.

The particular model to be developed is derived from the Hearsay-II speech understanding computer system. Hearsay-II has several features that make it at least a plausibly appropriate progenitor for a design model. These will be considered in turn.

Firstly, there are some similarities between the problems of speech understanding and design. Perhaps most importantly, they are both large problems with a high degree of solution uncertainty, requiring the coordination of a variety of different classes of knowledge to solve them. Thus understanding speech requires knowledge of syntax, semantics, phonetics, prosody, pragmatics, and so on. Similarly, designing a calculator, for example, might require knowledge of electronics, plastic mouldings, displays, fixing methods, mathematics, user tasks, and so on.

Another similarity is that problems in both areas allow more than one solution. Thus speech utterances can often be interpreted in more than one way. This is even more true of design problems, where the trade-offs between alternative solutions can be very fine. A further similarity is that solutions in both areas can be described at different levels. Words, syllables and phonemes are only three levels of description of a speech utterance. Design solutions might be described, for example, in terms of manufacturing drawings, general arrangement drawings and conceptual sketches.

Secondly, the way Hearsay-II operates has some interesting parallels to the psychological evidence on design. Hearsay-II is organised around what is called a 'hypothesize and test' paradigm. That is, hypotheses about the identities of portions of the speech signal are suggested and their plausibility is tested. This has obvious parallels to the evidence that designers formulate and test solutions, i.e. that they are solution-focused (Lawson, 1979). Another similarity is with the flexibility of control to be found during designing. Thus Hearsay-II uses a mixture of top-down and bottom-up processing to direct its solution search in a flexible manner. A further parallel can be found in the way that Hearsay-II constructs solutions incrementally, much as the evidence suggests designers do.

Thirdly, an interesting feature of the Hearsay-II architecture is that it was not conceived as a single fixed form suitable only for speech understanding. Rather it was seen as a framework for testing ideas about knowledge representation and cooperation. Thus Hearsay-II distinguishes a general framework from specific models. The framework provides a common architecture and is based on certain explicit assumptions; the models share the framework's architecture and assumptions but differ in the details of their content and operation. This distinction is a promising reason to think the comparison of design with and without CAD will be both possible and productive using this class of model. Different models for each type of design could be developed within the same framework. The distinction between a framework and a model has also been made within the psychological literature (e.g. Morton, Hammersley and Bekerian, 1985).

The final point in Hearsay-II's favour is that it works, and systems derived from it have proved successful in a number of different areas (see Hayes-Roth, 1983). This includes planning (Hayes-Roth and Hayes-Roth, 1979) which at least on the surface seems closely related to design.

Thus Hearsay-II has both structural and operational similarities with what is known about design, there is good reason to believe it will serve one of the model's main functions, and it is in some sense proven. Altogether, these provide good reason for taking Hearsay-II as a starting point for developing a model of design.

3.3.2   Description of Hearsay-II

A number of papers concerned with different aspects of Hearsay-II have been published. Among them, a full description appears in Erman, Hayes-Roth, Lesser and Reddy (1980), and a shorter one in Erman and Lesser (1980).

The Hearsay-II architecture is an attempt to solve the problem of communication between multiple classes of knowledge. It consists of a set of independent knowledge sources (KSs) that communicate with a central 'blackboard'. The blackboard is essentially a data structure divided into levels, each of which is a different representation of the problem space. It contains hypotheses as to the identities of portions of the speech signal. The KSs are kept separate and can communicate only with the central blackboard, not with each other. They can either create new hypotheses on the blackboard, or they can test the plausibility of existing hypotheses. This is known as the hypothesize and test paradigm. The blackboard organisation maintains a high level of flexibility of operation while minimising the representations and transformations involved.

The blackboard contains the current state of the problem solution, in the form of hypotheses about the identities of parts of the speech signal. It has three dimensions: time within the utterance, alternative competing hypotheses, and information level. The first of these is self-explanatory. The second reflects the fact that Hearsay-II pursues several possible interpretations in parallel. Different hypotheses at the same level and significantly overlapping in time are alternatives, and at some point in the understanding process a choice will be made between them.

The third dimension of information level is the most complex. The blackboard is partitioned into six levels: parameter, segment, syllable, word, word-sequence, and phrase. Each of these levels constitutes a different representation of the problem space, and each can hold a complete description of the utterance. What differentiates the levels is the unit making up its representation, e.g. word or phoneme. Hypotheses at all levels have a uniform attribute-value structure. Hypotheses at different levels are connected, usually by multiple links both above and below. This indicates the lower level evidence on which higher level hypotheses are based.

The Hearsay-II configuration normally discussed contains twelve KSs. Although these vary considerably in complexity, a KS can be schematized as a production rule of the form IF X THEN Y. It comprises two major components: a precondition and an action. The KS scans the blackboard to see if the appropriate conditions for action exist, and if so, attempts to implement its action. Its activation is thus not control-driven but data-driven, based on the occurrence of blackboard hypotheses forming appropriate conditions for action.

Hearsay-II essentially operates in a manner similar to that of a production system, in that it works in a series of recognise-act cycles. A cycle consists of all KSs scanning the blackboard to see if the conditions exist for them to act. Of those KSs that do wish to act, one is chosen and this carries out its action on the blackboard. Then a new cycle begins and all the KSs scan the blackboard again. The important work here is in the choice of which KS is allowed to act. This is done by a scheduler that calculates a priority for each KS action and selects the one with the highest priority for execution. The scheduler consists of a long and complicated formula designed to fulfil five principles, called the competition, validity, significance, efficiency and goal satisfaction principles.

3.3.3   The framework for the model of design

Given the architecture of Hearsay-II just described, it can be adapted, in line with the purposes and evidence outlined above, to act as a design model. This adapted framework is described here. The details of the model within this framework will be determined by the observational study to be described in the next section.

The framework consists of a set of knowledge sources accessing a central blackboard, on which the design solution is constructed. The blackboard has two dimensions – information level and spatial location. The framework merely states that there is at least one information level. Different models may vary in the number of levels they contain. The spatial location dimension replaces the time dimension of Hearsay-II. That is, in the way that parts of the speech utterance occur at certain times and can be so identified, parts of the design solution occur in identifiable spatial locations. The third dimension of Hearsay-II, alternative solutions, has been dropped. Although one can see parallel processing of alternatives as appropriate for an engineering solution to the problem of speech perception, none of the available evidence suggested designers work in that manner. Until the data suggest otherwise it is only parsimonious to assume this third dimension is not necessary for a design model.

A KS is an independent collection of knowledge pertaining to some aspect of the design problem (some detailed examples will be given in the next section). It operates in effect like a complex production rule. The condition component checks the blackboard for conditions appropriate to its action. If these are found, the action component attempts to implement its action on the blackboard. This is either to generate a new part of the solution, or to alter an existing part. Neither the numbers nor the identities of the KSs are specified in the framework.

The scheduling formula of Hearsay-II has been replaced. The formula is Hearsay-II's response to the control problem (i.e. determining which KS should be allowed to act when more than one is able to do so). Several systems in recent years have attempted to use some form of meta-knowledge to deal with the control problem. Meta-knowledge is knowledge about knowledge, i.e. it concerns some aspect of what one knows rather than of things in the world. It thus distinguishes object-level knowledge of the domain of interest (e.g. which syllable combinations form which words) from meta-level knowledge about the object-level knowledge (e.g. whether next to hypothesize new syllables or new words). Following Davis (1980), the design framework uses the idea of meta-level KSs for control in place of the scheduling formula. These are represented and applied in the same way as the object-level KSs. They do not write to the blackboard themselves, but schedule the application of the object-level KSs. This mechanism has advantages over the scheduling formula in that it allows the control knowledge to be represented explicitly and to be reasoned about. Consequently, the framework has a higher degree of flexibility. It can contain different strategies and tactics and choose between them according to the current features of the problem.

The framework as outlined here incorporates changes of three kinds from the original version of Hearsay-II. Firstly, there are changes appropriate for the different problem domain. These include altering the time dimension of the blackboard to a spatial location dimension, and dropping the third dimension of alternative solutions. Secondly, at least one change relates to a weakness of Hearsay-II, the scheduling formula. The use of meta-level KSs for control is an attempt to strengthen the framework. Thirdly, there are changes resulting from the different purposes of the two models. The purpose of Hearsay-II was to produce a program with certain performance characteristics. The purpose of the design model is to describe the design activity in a way that allows for a discussion of the nature and effects of computer aids. This means some detailed features of Hearsay-II necessary for its implementation are not relevant to the design framework.

For the reasons given earlier, this framework ought to be appropriate for developing models of design. How to do this, and the models that result, are the topics of the next two sections.

3.4   Development of the blackboard model

This section describes the observational study and data analyses that provided the content of the blackboard design models. The details of the models themselves will be presented in Section 3.5.

3.4.1   Methodology

Two groups of designers were studied, each of four subjects. One group worked on a CAD system and the other on conventional drawing boards. The CAD users will be termed the group of aided designers and referred to as GA. The drawing board users will be termed the group of unaided designers and referred to as GU.

The study is not strictly experimental in the conventional laboratory sense, since it lacks both the requisite control conditions and the random assignment of subjects to groups. But it does attempt to achieve a high level of ecological validity while still exerting a reasonable degree of control over the subjects' situation.

Each subject tackled one problem over a period of about two hours. The problems were allocated to the four subjects in each group on an ABBA basis. The two problems were devised specifically for this study. The first involves the design of a casing for a television monitor. The internal components are a cathode-ray tube (CRT), a printed circuit board (PCB) and a power supply. The sizes of these are given. Designing the casing includes arranging these internal components. The second problem involves designing the casing for the keyboard of a visual display unit, given a particular subassembly with the keys mounted on the PCB. Although this includes only one internal component compared to three in the first problem, its given dimensions do pose a particular difficulty of how to support the keyboard in the casing. Both problems were closely based on existing equipment.

There were three important considerations in the development of the problems:

- they should be at an intermediate level in the design process. This is what Pahl and Beitz (1984) call embodiment design. It contrasts on the one hand with conceptual design and on the other with detail design. Although most CAD systems focus on detail design, current developments aim to push them more towards embodiment design.

- the problems should be closely related to the subjects' area of work, as part of the attempt to attain a high level of ecological validity.

- the problems should be of such a size that a reasonable attempt could be made at them over two hours. The aim was to get one substantially complete general arrangement drawing showing all the components and their interrelationships. Two hours is a shorter period than would normally be spent on problems of this magnitude, and the potential effects of this should be borne in mind.

All the subjects were practising mechanical engineering designers working in industry. As is usual in this profession, all were male. Subject group GU worked for the instrumentation group of a large electronics company. The major part of their job was to design instrument housings, involving both free-standing and rack mounted units. The ages of the subjects averaged 37 years, ranging from 25 to 53. They all had some form of qualification in mechanical engineering. Their experience in the field of design and draughting was between 5 and 21 years (mean 13). These subjects did all their work on drawing boards. They had very recently undertaken a short introductory training programme for CAD but did not yet use it for production work.

Subject group GA worked for the instrumentation division of a large manufacturing company. Most of the subjects' work involved the design of automotive instrument panel layouts and housings. The ages of the four subjects averaged 38 years (range 23 to 51). They had a mean of 14 years experience of design and draughting (range 1 to 25). Three of the subjects had qualifications in mechanical engineering, while the fourth was qualified in industrial design. The company had used a CAD system for several years. The experience of the individual subjects with the system ranged from 1 to 4 years (mean 3). The equipment they used was a standard system from one of the world's leading CAD vendors, and can therefore be considered reasonably representative of those available.

The two subject groups are therefore very similar in terms of age and design experience, and to a lesser extent, qualifications. They are also engaged upon similar work, i.e. the design of instrument casings, although there are bound to be differences in the actual work details.

The recording equipment provided both video and audio records, plus timing information (accurate to 1/10th second). For a combination of technical and methodological reasons the video camera was positioned to the side of the subject and maintained the subject within the frame. This means details of individual lines drawn cannot be distinguished on the recording.

All subjects were tested individually at their place of work. The instructions to the subjects asked them to provide a running commentary of what they were thinking as they did the task. The observer tried to restrict his utterances to requests for comment or clarification. The session terminated at a suitable point after two hours.

3.4.2 Data analysis

The principal data obtained are the verbal protocols (referred to from now on simply as protocols). These are supplemented by the video recordings and by the drawings and sketches. Firstly, each complete protocol was transcribed. Then two protocols from each of groups GU and GA were selected for detailed analysis. It was felt that this would be sufficient to represent the data fully and this was confirmed by subsequent partial analysis of the remaining protocols. The selection was done on the basis that: each problem was covered in each group; the shortest protocol in each group was excluded; and the subjects did not appear to adopt unusual or idiosyncratic methods. These requirements were not found to conflict. The four protocols selected were those of subjects 1 and 2 in GU, and subjects 2 and 4 in GA. These will be referred to as S1U, S2U, S6A and S8A respectively.

Each of the four protocols was then exhaustively classified twice by the author. The first classification was to assign each utterance to one of five categories. The second classification was according to a hierarchic problem breakdown. This two-part analysis was devised as a means by which the framework could be used to model the data. The derivations of the two classifications will be given along with their details, as follows.

The first classification assigned utterances to one or more of five categories:

- *generation* of new hypotheses
- *evaluation* of existing hypotheses
- *reasoning* behind hypothesis generation or evaluation
- *statement* of conditions or existing hypotheses
- *meta-statement* about control

From the protocol of one subject (S2U), example utterances in each of these categories are:

Generation: I would expect to see the printed circuit board sitting at the bottom there. (at time (hour:min:sec) 0:11:26)

So we'll need the other one far enough away, perhaps about 60 millimetres apart. And then again, set it up from the bottom, probably about 40. (1:02:00)

Evaluation: Be quite easy for the maintenance man to get at, I think, which is another thing. (1:35:05)

And if you're going moulding it's not going to cost – a little bit more for tooling perhaps to have a slight radius rather than straight sides. But not enough to give us any worries. (0:18:37)

Reasoning: Well, we're talking about the home computer market, so therefore I'm assuming we're talking of large volume. If we were talking about very small volume, in the order of twenty to thirty, the sort of thing we do here, twenty or thirty a year, then some sort of sheet metal would be obviously the most economical thing. (0:06:53)

Statement: We've got a no-go area of three millimetres underneath. And the components are 25 the other side. (0:54:45)

Meta-statement: The biggest problem as I see it, apart from the shape of the box or whatever we're going to put it in, the biggest problem is mounting the CRT. Mounting the power supply and mounting the PCB is no problem at all. (0:04:47)

I think that would be a small price to pay for making the thing look a bit better on the front. So we'll settle for that one. (0:29:14)

I think I'll have a look at the back view and see how it looks, with a view to moving that board over. (1:26:32)

The division of the protocol into individual utterances was not grammatically constrained. Thus an utterance can be of any length. There were no formal criteria for dividing the protocol. Informal cues included pauses, topic changes, interruptions of or interpolations into sentences, and the use of particular words or phrases. Utterances were classified into more than one category if there was uncertainty about which was the correct classification.

What might be called the arguments of the generation and evaluation utterances are taken to constitute the individual KSs. Thus the examples of generative utterances given above would be taken to indicate the presence of a KS concerned with parts arrangement. Similarly, the evaluative utterances above would be taken as evidence for KSs dealing with maintenance and cost. The number of KS distinctions that it is reasonable to make here is to some extent a matter of judgement. Too few means different activities will not be discriminated and the model will lack power. Too many means the grounds for identifying different activities will be unreliable.

Each utterance category provides a different class of information for the model: generation utterances indicate KSs which create new hypotheses either within or between blackboard levels; evaluation utterances indicate within-level KSs which modify existing hypotheses; reasoning utterances illustrate part of a KS's content; statements describe the contents of the blackboard; and meta-statements indicate the control knowledge.

This classification was derived from the combination of a number of distinctions. That between generation and evaluation arises from the design literature; its correspondence with the creation and modification of blackboard hypotheses makes it appropriate for this framework. The distinction between object knowledge and meta-knowledge arises from the framework itself; the distinction must be made to develop the model within the framework. The discrimination of reasoning and statement utterances arises from relating the data to the framework; the protocols contain both explanations of some KS applications and descriptions of the current design solution. The classification is thus driven in large part by assumptions about the suitability of the framework.

The classification is not absolute either in the sense that it is the only one suitable for this task, or in the sense that everyone using it would inevitably arrive at the same result. Of course it ought to be validated by measuring the agreement in its use by different judges. Assignment of utterances to categories is a matter of judgement, although the ability to assign them to more than one category is an attempt to negate individual biases in this respect. The low percentage of utterances that were unclassifiable (2%) or assigned more than once (5%) is evidence that the category divisions are reasonably sound.

In terms of the model, this first classification: identifies which KSs were used in solving these problems; specifies at least some of the content of those KSs; and both partly specifies and allows one to infer the control knowledge utilised by the subjects. It is not informative about the number of levels to the blackboard, or about the levels at which each KS acts. These were the aims of the second classification. The need for this classification therefore arose from the requirement to complete the model details not met by the first classification.

The second classification divided each protocol according to what was inferred to be the current problem on which the subject was working. This involved a hierarchic problem breakdown. This breakdown was not predetermined, either in terms of levels or content, but arose from the combination of what the subject said and the researcher's understanding of the problem. Such a hierarchic view of design problems is in line with the psychological evidence discussed in Section 3.2.

When, in the first classification, a KS appeared in a protocol, it was taken as acting at the level on the blackboard indicated by the contemporaneous level in the second classification. This classification for all protocols resulted in only a three-level problem breakdown. The models therefore contain three blackboard levels. These were called the *unit, item* and *detail* levels. Had the classifications resulted in more or less levels than three, then the blackboards in the models would have reflected this. The number of levels identified here is, as with the first classification, a matter of judgement. Too few or too many levels run the risks of a lack of power and unreliability respectively. With only three levels, this classification clearly trades off power for reliability. This means that any lack of discrimination between the subjects in terms of this classification cannot be accounted for solely by similarities between the subjects; it may also reflect a lack of power in the methodology.

The combination of these two classifications allows the protocols to be analysed in a way that provides the details of individual models in terms of the blackboard framework. These details will be described in the next section.

The video recordings served to disambiguate some protocol utterances and allowed all of them to be uniquely identified by time. They were also used to divide the subjects' performance into drawing and non-drawing periods. This analysis relates to the comparison of aided and unaided design (see Whitefield 1986a) and will not be discussed here. The subjects' drawings and sketches were not themselves analysed, but were used selectively to illustrate points arising from the models and their comparison. The subjects also completed a short questionnaire asking for their views on the session. In general they all felt that the problems were suitable, and that they had worked fairly normally, with no negative effects from having to provide a concurrent protocol.

3.5   Content of the blackboard model

This section describes the models of design arising from the observational study outlined in the last section. For space and convenience, the description will concentrate on the GU model, with the GA model considered only briefly.

3.5.1   *The KSs*

Table 3.1 lists the KSs identified in each of the four fully analysed protocols. The KSs are grouped according to two important and orthogonal

	S1U	S2U	S6A	S8A
GENERATIVE DOMAIN KSs	materials parts arrangement casing parts fixings wiring connections unspecified parts details	materials parts arrangement casing parts fixings wiring connections unspecified parts details	materials parts arrangement casing parts fixings wiring connections unspecified parts details	materials parts arrangement casing parts fixings wiring connections
EVALUATIVE DOMAIN KSs	support size assembly space utilisation appearance cost moulding production maintenance safety environmental hazards related equipment	support size assembly space utilisation appearance cost moulding production maintenance usability ventilation market acceptance	support size assembly space utilisation usability	support size assembly space utilisation appearance cost moulding production
TOTAL DOMAIN KSs	17	17	11	12
GENERATIVE DRAWING KSs	production management	production management	production management navigation	production management navigation
EVALUATIVE DRAWING KSs	interpretation	interpretation	interpretation	interpretation
TOTAL DRAWING KSs	3	3	4	4
TOTAL KSs	20	20	15	16

Table 3.1: KSs evaluated in each protocol

distinctions. That between generative and evaluative knowledge was covered in Section 3.4. The other is between those KSs containing knowledge about the domain (i.e. knowledge of casings) and those containing knowledge about drawing. One can see from the table that each of the two pairs of subjects is very similar and that there are differences between the aided and unaided subjects. The two unaided subjects each produced 20 KSs, of which 17 were common to both, making a total of 23 KSs between them. The two aided subjects produced 15 and 16 KSs each, of which 13 were common to both, making a total of 18 KSs between them. It is worth noting that all the KSs produced by GA were also produced by GU, with the single exception of drawing navigation. Some of the KSs will be described briefly here, with example utterances from the protocols. The purpose is to illustrate the types of KSs, their concerns, their application, and how they appear in the protocols. The domain KSs will be considered first.

The generative domain KSs have the role of suggesting on the blackboard possible solution hypotheses. These might serve as a basis for other generative KSs to suggest further hypotheses, or for critical consideration by evaluative KSs.

Fixings: generates methods for mounting the internal components to the case, and connecting the parts of the case to each other. This KS includes a wide variety of different methods, such as screws, bolts, snap-fits, hinges, clamps, slot-fits, heat bonding, and so on. The hypotheses suggested can be general or specific ("I would expect to see some sort of metal clamping strip, probably in two halves, going round there, clamping through. Yes, to clamp the CRT" S2U, 0:10:27; "I should think something like an M4 screw would do there" S1U, 1:06:11.)

Casing parts: generates the individual sections of the casing, both in number and in size and shape. ("It's possible I suppose we could have a front moulding, which would be something like that. And then a second moulding, which probably could be the body of the thing, and just leaving us the facility to put a loose cover in the back" S2U, 0:38:22; "If we set a nominal thickness of the front panel at 3mm" S2U, 1:13:12.)

Materials: is used in choosing the materials from which to make the case. This is normally evoked early on, since the decision has consequences for the way the object is designed ("I would start off by assuming that we're going to have some sort of plastic moulded case for the thing" S2U, 0:06:43). If the case were to be made of sheet metal instead, this would affect the parts that make up the case and the fixing methods used. The same KS would be used at a later point in deciding exactly which kind of plastic to use.

The evaluative domain KSs serve to criticise hypotheses placed on the blackboard by the generative KSs. A negative criticism can result in a hypothesis being downgraded or even removed from the blackboard, necessitating that the generative KSs produce another. A positive criticism produces more confidence in a hypothesis. The relative importance assigned to these different KSs will vary according to both the problem specification and the designer's individual preferences.

Appearance: parts of the design are evaluated on appearance grounds. This can be based on non-specific personal preferences, or on more specific criteria. (On having screws showing on the top surface: "I don't want to do that. That's not going to look good at all" S1U, 0:22:13; "I was just wondering whether that might not be an enhancement, you know, a nice extrusion around the keys" S1U, 0:36:10.)

Space utilisation: this is concerned both with conflicts between parts for the same spatial location and with how much wasted space the casing encloses. An efficient spatial arrangement of the parts has positive consequences for both size and materials, especially for the monitor problem. ("It may work out that where I want to put my web is where the slot's going to come" S1U, 2:04:10.)

Support: evaluates the stability of the whole or parts. This is particularly relevant for criticising fixing methods. ("the order of five or six millimetres. That should be man enough for that, to take that" S2U, 1:21:50).

Maintenance: various factors are important for maintenance, particularly the accessibility of the parts and repeated use of the fixings. ("Although with self-tappers, there is this problem that, for maintenance, you know, withdrawing and inserting these things a few times, you're in trouble" S1U, 0:30:59.)

Usability: perhaps more rarely than ergonomists might like, the subjects do occasionally evoke knowledge of usability in evaluating solutions. ("I suppose they ought to go on the front. Because that's where people will be" S2U, 0:26:12).

Generative drawing KSs perform the job of translating the mental blackboard representations into physical drawings and sketches. The GU subjects used two kinds of knowledge for this.

Drawing production: responsible for producing both freehand sketches, at the item level, and formal drawings, at the detail level. It contains knowledge of how to represent and draw parts in traditional orthographic and perspective projections.

Drawing management: as well as actually drawing the lines, the designer needs to decide the views to be shown, their arrangement on the sheet, the scale of the drawing, and so on. For GU, use of this KS tends to be concentrated in a short period around the time when the drawing is begun. Because of the high overheads of redrawing from scratch, it is important that this knowledge is applied correctly. ("I'm just seeing what sort of space I need to fit it in. I don't have to draw the complete length obviously, it's the end I'm interested in, and round that area there. So I can do broken views to show the things I want" S1U, 0:59:10.)

Only one evaluative drawing KS was found in the protocols.

Drawing interpretation: the drawings and sketches can be considered as real-world embodiments of the detail and item levels of the blackboard. The KSs can use them as input in the same way as using the black-board. Occasionally the subjects need to identify or to work out exactly what certain lines represent. ("OK, I've drawn it flat as opposed to on a slope" S1U, 0:30:29; "No, that's not right what I've drawn there, is it?" S1U, 1:14:16.)

The problem classification of the protocols allows the domain KSs to be allocated to blackboard levels. As an example, Figure 3.1 shows the relationships of the domain KSs to the blackboard for subject S1U. The three blackboard information levels are shown only as lines, although each line represents the second blackboard dimension of spatial location within the design. Thus solution parts at each level occur in identifiable locations in three-dimensional space. The sketch and drawing boxes represent the freehand sketches and technical drawings produced by the designer. They are shown as continuous with the item and detail levels of the blackboard since, as stated above, they can be considered as real-world embodiments of those levels and can be used by KSs as input in the same way. The circles represent the level of input to a KS, and the arrows its level of output. KS types A and B are generative and can operate either within or between levels. Types C and D are evaluative and operate only within a level. A list of the KSs of each type accompanies the figure. Those KSs given in capitals were found under two levels in the problem classification, and therefore appear as more than one type.

In general, the subjects applied most of the generative KSs at both the item and detail levels of the blackboard. But a smaller proportion of the evaluative KSs were used at the detail level. This is clearly the case for S1U in Figure 3.1. The general trend implies that the generation of solutions occurs at all levels, while they are more likely to be subjected to the full range of evaluations at the item level.

3.5.2 Control and operation

The data analyses provide no scheme for further treatment of the meta-level control knowledge after its classification as such. An informal analysis of the meta-level utterances suggests they are concerned with the following five aspects of the task:

- ordering activities (e.g. deciding which sub-problem to tackle next)
- prioritising evaluations (e.g. choosing between conflicting evaluations)
- sub-dividing the problem and assessing the difficulty of each part
- reporting high-level summaries of object-level activity
- identifying missing knowledge that would help in constructing a solution (e.g. that one would need to talk to a production engineer before making a certain decision)

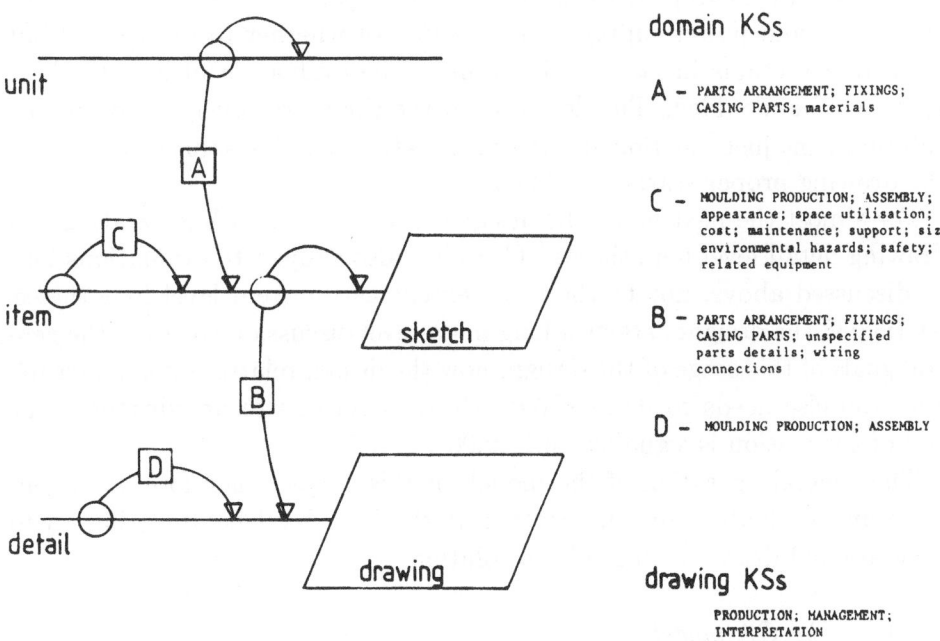

Figure 3.1: Allocation of domain KSs to blackboard levels for subject S1U

Taking as an example the protocol of S2U on the monitor problem, the operation of the model described here would be as follows. Certain items from the problem statement are placed on the unit level of the blackboard. These would be the internal components and a casing. The additional information in the problem statement allows the designer to adjust the priorities of relevant KSs. The designer also divides the problem into a number of sub-problems that are fairly independent, and decides which to tackle first. These activities are all present within the first seven minutes of the protocol.

At this stage, the generative domain KSs produce new hypotheses at the item level. The materials KS is evoked and decides what the case should be made of. This is followed by a number of applications of the parts arrangement, casing parts and fixings KSs, which generate a rough solution that is also drawn as a sketch. Once there is sufficient content at the item level, the hypotheses there begin to be criticised by the evaluative domain KSs (from c.0:12:00).

This interplay between generation and evaluation KSs continues until a satisfactory approximate solution is built up at the item level. Then (c.0:33:00) the designer starts considering the problems of drawing management. By this point he has covered all the sub-problems in the first level breakdown except dividing the case into parts and mounting the circuit board and power supply. The question of whether to draw small but important sections in large scale prompts the evaluation of how the case parts will be moulded. This leads on to the three previously unconsidered sub-problems just mentioned. After generating possible solutions to these, the drawing proper starts (c.0:43:00).

For almost the next hour, the designer is concerned with generating and drawing detail level hypotheses. These are also subject to evaluations but, as discussed above, not to the same extent as the item level hypotheses. At 1:38:38 the designer starts a long meta-level discussion covering the general goals of this stage of the design, how the design relates to manufacture, and who else needs to be involved. He then returns to drawing until the end of the session is signalled (c.1:59:00).

The general operation of the model for this subject therefore, is to generate and evaluate a full description at the item level before going on to generate and draw the detail level solution.

3.5.3   *The GA model*

The description of the GU model just given will also serve here to characterise the GA model. The differences between the two (some of which are evident from the table above) are not important for the purposes of this chapter. Suffice to say the framework is capable of modelling two types of design in a way that allows a range of similarities and differences to be identified. These concern the identities, numbers and contents of KSs, their frequency of use, their allocation to blackboard levels, and the control knowledge that applies them. A comparison of the two models is presented in Whitefield (1986a).

3.6   Application of the model

This chapter concentrates on the model's construction. But it is impossible to assess this (the topic of the next section) without some knowledge of its application. As described in Section 3.3, the principal intended application of the model is to assist CAD system developers in structuring CAD programs appropriately for their users. Applying human-computer interaction (HCI) models to system development is a relatively novel activity. Reisner describes it as "an area which is still groping for basic concepts" (1983, p.117). There are therefore no accepted procedures for it and the work done so far has been fairly disparate.

Whitefield (1986b) attempted to distinguish the different kinds of help which the HCI models currently in the literature offer developers. Four model roles in particular were outlined and compared. The role of the blackboard model presented here is to assess user/program compatibility. It does this by allowing the developer to reason about the consequences of possible program features on the user's knowledge recruitment. The application of the model is certainly not routine and it demands a considerable investment of time and effort from the developer.

The model has not yet been used by real system developers. A simulated attempt to apply it and hence to demonstrate its usefulness is described in Whitefield (1986c). Briefly, this consisted of devising two versions of a small CAD program which differed in their compatibility with the user as predicted by the model. Task performance of users with the two programs was then experimentally compared. The prediction was that the higher compatibility system would lead to better performance. A number of measures suggested the predicted outcomes did in fact occur.

The important point here of this simulated application of the model is how the two program versions were devised and assessed using the blackboard model. The first problem was deciding the criteria to assess compatibility. To do this, one needs a view of what constitutes a desirable use of knowledge by the designer. The view adopted derives from the definition of design and the evidence on design covered in Section 3.2, and from the comparison of aided and unaided design. The criteria follow from this and were expressed in terms of encouraging or discouraging the application of certain classes of KS (e.g. domain knowledge).

Given suitable criteria, one needs a means of representing the proposed program so it can be matched with the model. The simulated exercise involved altering an existing program and there were severe constraints on how much could be altered. The representation of the program used for this was the command language. The kinds of changes implemented involved prompting domain evaluations, reducing the navigation requirement, and alternative forms for libraries of parts. The command language representation would almost certainly be unsuitable for earlier stages in system development, and current work is investigating the use of alternative representations. In any case, the representation should allow the developer to consider possible system features and, in conjunction with the blackboard model, to reason about their effects on the user's knowledge recruitment. The outcome of this would be an analysis of how particular features support task performance in terms of the relevant criteria.

For the model to be usable by system developers many improvements would be desirable. Certainly a greater clarification of the mechanics of its application, and possibly an alternative form for its presentation, would be two improvements. But the simulation exercise has shown both what is involved in the model's use (e.g. the kinds of representations and criteria needed) and that the kinds of changes it can assess do lead to performance improvements. It therefore provides a promising start to demonstrating the model's usefulness.

3.7   Assessment

This chapter has attempted to relate a case study of the construction of an appropriate model of a user to be applied during computer system development. It might be as well at this point to summarise the main steps in this process. Further steps would be involved in any application of the model but these have not been the focus of this chapter.

1. Assess the evidence on the user's task; this involves selecting the relevant material from the available empirical evidence.
2. Select a suitable framework; this should be suitable for both the user's task (about which one has just discovered) and the system developer's task (about which we currently know little).
3. Adapt the framework; there will inevitably be unsuitable features of any extant framework which should be changed for the current purposes.
4. Detail the model empirically; some form of direct evidence is necessary to provide the real content of the model.

This process can be assessed in at least two ways, which will be considered in turn. The first concerns the adequacy of the particular framework and model described. The second concerns the approach this work embodies towards ergonomic research and application.

Regarding the particular process, its strengths lie in the early steps (the selection of a suitable form and the framework/model distinction) while its weaknesses are apparent later on (the empirical methodology) and these will be considered in turn. The blackboard framework is a suitable form for a number of reasons. The similarities between Hearsay-II and the phenomena of design were discussed in Section 3.3. In addition, the nature of the framework as a knowledge description is an advantage. Its orientation towards the designer's knowledge matches the status of design as a predominantly cognitive activity. As a description, the blackboard models are concerned with the classes of knowledge utilised and an architecture within which their application can be understood. In contrast, most design models are prescriptive (see Lawson, 1980). They prescribe an ordered list of phases or of the products of those phases. The problem with prescriptions for an activity as variable as design is that they mostly fail to accord with the behavioural evidence. If one is to relate possible CAD systems to models of design, descriptive models are likely to be both more accurate sources and more flexible in terms of what they can be related to.

The distinction between a framework and a model has proved useful in facilitating both the development of the models and the comparison between them. It has done the former by two sorts of clarification. The first is between the general and specific features of the model (e.g. that knowledge is represented in KSs versus the particular KSs used). The second (and overlapping) clarification is between those aspects of the models deriving from the data (e.g. the contents of any KS) and those deriving from other considerations (e.g. the suitability of a blackboard architecture). Some justification for the particular framework chosen is it has proved capable of application to a rich and complex engineering design task. In addition, it has subsequently been successfully applied to tasks in the design of steam turbine pipework (Warren and Whitefield, 1987). The framework's derivation and its success with two different sorts of engineering design task suggest it could be used for a wide range of such tasks.

The framework/model distinction has also facilitated the comparison between the models (Whitefield, 1986a). This comparability is an important asset. Since both the GU and GA models are developed within the same framework, the grounds for their comparison are clear. There is no reason to believe the framework is more appropriate for either aided or unaided design. Without this common derivation the grounds for comparing models would be much more uncertain.

The weaknesses of the approach largely stem from the methodology used to detail the models empirically. This has different weaknesses from each of two perspectives. From a scientific point of view the models rely too heavily on a small number of subjects, on only two problems, and on verbal protocols as a data source. Attempting to remedy these deficiencies would be an important part of any future work to improve the scientific status of the models. From an applications perspective the weaknesses of the methodology concern the time and resources used in the model construction. Under the usual time pressures involved in writing software, few system developers could afford this level of commitment. One defence here is that, being exploratory, this work is unlikely to have been economical. A second is that one could now streamline the construction process so it requires the minimum of resources to produce the same outcome. This is one aim of current work.

Of course, some weaknesses are seen from both the scientific and the applications perspectives. As mentioned above, the protocol analysis suffers from being unvalidated. Both classifications rely on subjective judgements to an extent that ideally requires a measure of agreement between different judges on the reliability of category assignments. One of the advantages of the analysis, however, is it does provide a complete treatment of the protocols, with all utterances classified. This completeness ensures data are not missed. Allied to the fact that there is little within-group variation, this suggests the models would not alter much if more subjects were observed.

In spite of their exhaustive analysis, the protocols are unable to provide a detailed description of all the actual knowledge used, although it is not clear this would be appropriate for the applications of the models. The designers do not verbalise everything they think, and what they say may not always be an accurate reflection of what they think and do. This incompleteness and inaccuracy mean a description of all the knowledge applied would be impossible to obtain from such a study. Listing the knowledge classes, therefore, is likely to be a more accurate description of the activity, relative to its own (fairly high) level, than is a more detailed description of the knowledge.

In sum, therefore, the advantages of the process described here are in the the selection of an appropriate form, and the distinction between the framework and the model. The major disadvantages arise from the empirical methodology employed.

As an instance of ergonomic research, this work of course takes a particular approach to the problem under study. Given the book in which this chapter appears, it is appropriate here to use Long's framework for ergonomics (Long, 1987; this volume) to characterise this approach. The interesting features of it can then be discussed.

In brief, Long presents a framework for describing ergonomic activities in terms of particular configurations of tasks and sciences. It consists of a set of representations and transformations. Starting with tasks in the real world, analysis produces an acquisition representation which supports laboratory simulation for experimentation. The scientific knowledge representation is a result of generalisation from the experimental data. This in turn is particularised to produce an applications representation, such as design guidelines. The final transformation is to synthesize this applications representation with tasks in the real world to improve the human/machine relationship.

This framework describes the work in this chapter as follows. An analysis of the real world design task selects that version of it which constitutes the acquisition representation. This is the basis of the observational study. The representation selected remains close to the real world of work (i.e. aims for high ecological validity) in that three of the task features (agents, instruments and location) are identical, with only minor alterations to the other two (functions and entities). The protocol database arising from the simulation is then combined with a representation from the science base (Hearsay-II) to form the blackboard model. Where this representation sits in the ergonomics framework relative to the science base will be discussed below. Without further transformation, the blackboard model constitutes a proto-applications representation. That is, it has not been directly used in a synthesis transformation by system developers, but its use has been explored in the simulated synthesis outlined in Section 3.6. How the model might be transformed into a full applications representation was also briefly mentioned there.

The particular configuration of tasks and sciences involved, and the task problems addressed, are what identify this work as an example of cognitive ergonomics. Thus it focuses on predominantly mental tasks, draws on an artificial intelligence representation, and addresses representational incompatibility problems concerned with task functions and entities. Two specifically interesting features of the work within this overall framework are the sequence of representations and transformations involved, and the relationship between the two main activities of acquiring and applying knowledge.

Concerning the sequence of representations and transformations, the work starts with an analysis of the engineering design task to produce the acquisition representation. In Long's framework the next transformations would be to generalise from the protocol database to a science knowledge representation, and then to particularise this to produce an applications representation. This work curtails that route by transforming the protocol database into the blackboard model, which is itself intended as the applications representation. Thus the acquisition representation is transformed directly into the applications representation without going through an intermediate science base representation. A contrasting example would be the work of Card, Moran and Newell (1983). Their database of text editing performance is generalised into what is called the GOMS model. This is a science base representation, intended as an information processing model of a routine cognitive skill. The GOMS model is then particularised into the Keystroke Level Model, which is the applications representation intended for use by computer system developers.

Of course the transformation from the protocols to the blackboard model draws on the science base (in the form of Hearsay-II) to develop the representation. But it does not contribute to the science base since the blackboard model is not being put forward as a psychological model of design. There are reasons for thinking it is a good candidate for development as a scientific psychological model, but in its current state it does not make that claim. As mentioned earlier in this section, widening the sets of subjects, tasks and data involved in this study would be a major part of developing the blackboard model as a scientific model.

Thus there is no separate representation in the science base. Clearly this route from knowledge acquisition to application attempts to be direct and might be considered more of an engineering than a scientific approach to ergonomics (see Long, 1986). Related to this is the work's other interesting feature – it covers both the two main ergonomic activities of acquiring knowledge about tasks in the world and applying this knowledge to improve human/machine relationships in tasks. The focus here has been on the former activity with the latter considered only briefly in Section 3.6. Although ergonomics has always consisted of these two activities, there is often a tendency for them to be conducted separately – the former in universities and research establishments and the latter by practitioners in industry. But if researchers are to have a positive impact on human-computer interaction, it is important that they attempt to formulate their research output with due regard to how it can be applied.

3.8   Summary

This chapter has described a case study involving the construction of a model of the engineering design activity that can be applied by CAD system developers to help in producing suitably structured programs. The model's construction began with an assessment of what is already known about how engineering designers carry out their task. This provided information that allowed the selection of an appropriate form for the model. The developer's task is another possible constraining influence on the choice of form. The Hearsay-II representation (from artificial intelligence) was selected for its parallels with design. It was adapted in the light of the new domain, the new purpose, and its specific weaknesses. An empirical study involving a realistic design task was conducted to provide the data for the detailed content of the model. The data analyses were developed to construct the detailed model within the general framework. With some criteria about what is good design practice, the model can be applied to reason about the consequences of different program structures on the user's knowledge recruitment, and hence on the suitability of the proposed program. This application was covered only briefly here, and has been carried out only in a simulated form. Although weak in its empirical database, the model is not intended as a science base representation but as a direct link between the analysis of the real world task and the synthesis of new interactive CAD systems.

Acknowledgements

I would like to thank the companies who allowed access to the designers, and my colleagues at the Ergonomics Unit who commented on an earlier version of the chapter. This work was done while I was in receipt of a research studentship from the Science and Engineering Research Council, supervised by John Long.

References

Adelson B., Littman D., Ehrlich K., Black J., and Soloway E. (1985): Novice-expert differences in software design. In: B. Shackel (ed): *Human-Computer Interaction – INTERACT 84*. Amsterdam: North-Holland.

Akin O. (1978): How do architects design? In: J.C. Latombe (ed): *Artificial Intelligence and Pattern Recognition in Computer Aided Design*. Amsterdam: North-Holland.

Card S.K., Moran T.P., and Newell A. (1983): *The Psychology of Human-Computer Interaction*. Hillsdale, New Jersey: Lawrence Erlbaum.

Cross N. (ed) (1984): *Developments in Design Methodology*. Chichester: John Wiley & Sons.

Davis R. (1980): Meta-rules: reasoning about control. *Artificial Intelligence*, **15**, 179-222.

Eastman C.M. (1970): On the analysis of intuitive design processes. In: G.T. Moore (ed): *Emerging Methods in Environmental Design and Planning*. Cambridge, Mass: MIT Press.

Erman L.D., Hayes-Roth F., Lesser V.R., and Reddy D.R. (1980): The Hearsay-II speech-understanding system: integrating knowledge to resolve uncertainty. *Computing Surveys*, **12**(2), 213-253.

Erman L.D. and Lesser V.R. (1980): The Hearsay-II speech understanding system: a tutorial. In W.A. Lea (ed): *Trends In Speech Recognition*. Englewood Cliffs, New Jersey: Prentice-Hall.

Hayes-Roth B. (1983): The blackboard architecture: a general framework for problem solving? *HPP Report* No. HPP-83-30, Stanford University, Computer Science Dept.

Hayes-Roth B. and Hayes-Roth F. (1979): A cognitive model of planning. *Cognitive Science*, **3**, 275-310.

Henrion M. (1978): Automatic space-planning: a postmortem? In: J.C. Latombe (ed): *Artificial Intelligence and Pattern Recognition in Computer Aided Design.* Amsterdam: North-Holland.

Jeffries R., Turner A.A., Polson P.G., and Atwood M.E. (1981): The processes involved in designing software. In: J.R. Anderson (ed): *Cognitive Skills and Their Acquisition.* Hillsdale, New Jersey: Lawrence Erlbaum.

Lawson B.R. (1979): Cognitive strategies in architectural design. *Ergonomics*, **22**, 59-68.

Lawson B.R. (1980): *How Designers Think.* London: Architectural Press.

Lera S. (1983): Synopses of some recent published studies of the design process and designer behaviour. *Design Studies*, **4**, 133-140.

Long J.B. (1986): "People And Computers: Designing For Usability" – an introduction to HCI'86. In: M.D. Harrison and A.F. Monk (eds): *People And Computers: Designing For Usability.* Cambridge: Cambridge University Press.

Long J.B. (1987): Cognitive ergonomics and human-computer interaction. In: P.B. Warr (ed): *Psychology At Work.* Third edition. Harmondsworth, Middx: Penguin.

Malhotra A., Thomas J.C., Carroll J.M., and Miller L.A. (1980): Cognitive processes in design. *International Journal of Man-Machine Studies*, **12**, 119-140.

Morton J., Hammersley R.H., and Bekerian D.A. (1985): Headed records: a model for memory and its failures. *Cognition*, **20**, 1-23.

Pahl G. and Beitz W. (1984): *Engineering Design.* English language edition, edited by K. Wallace. London: Design Council.

Reisner P. (1983): Analytic tools for human factors of software. In: A Blaser and M. Zoeppritz (eds): *Enduser Systems and their Human Factors.* Lecture Notes in Computer Science No. 150. Berlin: Springer-Verlag.

Stefik M., Aikins J., Balzer R., Benoit J., Birnbaum L., Hayes-Roth F., and Sacerdoti E. (1982): The organization of expert systems: a prescriptive tutorial. Xerox Parc Report VLSI-82-1.

Warren C.P. and Whitefield A.D. (1987): The role of task characterisation in transferring models of users: the example of engineering design. In: H-J. Bullinger and B. Shackel (eds): *Human-Computer Interaction – INTERACT'87.* Amsterdam: North-Holland.

Whitefield A.D. (1986a): An analysis and comparison of knowledge use in designing with and without CAD. In: *Knowledge Engineering And Computer Modelling In CAD – Proceedings of CAD86.* London: Butterworths.

Whitefield A.D. (1986b): Human-computer interaction models and their use in computer system design. In: *Proceedings of the Third European Conference on Cognitive Ergonomics – ECCE3.* Paris, September 1986.

Whitefield A.D. (1986c): *Constructing and applying a model of the user for computer system development: the case of computer aided design.* Unpublished PhD thesis, University of London.

4

Developing a science base for the naming of computer commands

Phil Barnard,

Jonathan Grudin,

Allan Maclean

4.1   Introduction

4.1.1   Creating and using a science base for human-computer interaction

The study of human-computer interaction seeks to understand how factors within some overall context of use contribute to the productive and efficient use of hardware and software (Barnard, 1988). The real world context of use involves people interacting with computer systems in order to complete some task. Clearly, numerous factors contribute to this context: the knowledge, experience, and physical properties of the user; the domain of application; the task in hand; the system functionality and its style of dialogue; the history of a particular dialogue exchange and so forth. Developing an understanding of how such factors affect computer use suggests two basic requirements. It requires a science base in the form of systematic knowledge of what governs user behaviour. It also requires a capability to relate that knowledge in a principled manner to some real world problem such as the design of systems or of training programmes.

In providing a framework for this book, Long makes two fundamental points. First, the kind of understanding embodied in the science base is itself a representation that can only be mapped to and from the real world via intermediary representations. Second, the precise representations and mappings required to realise this two-way conceptual traffic are dependent upon the activities they are required to support. Thus, the representations and mappings called upon in activities associated with basic science will differ from those involved with system design and development.

When operating within a scientific paradigm directed at developing an applicable understanding of behaviour in human-computer interaction, we make assumptions both about what to abstract from the real world and about the empirical and conceptual techniques that best capture this abstraction. Long refers to this abstraction as an "acquisition representation." As an intermediary representation, one of its functions is to support a simulation of the real world for the purposes of gathering data. Regularities in the data and any theoretical interpretation of them constitute the systematic knowledge embodied in the science base. Likewise, in order to apply that systematic knowledge, complementary assumptions are made concerning precisely what knowledge should be called upon from the science base and how it should be mapped onto the real world. This Long refers to as an "applications representation".

Within Long's framework, strategies for using acquisition and applications representations in the creation and exploitation of a science base can readily be described. A formalist, for example, might tackle a problem by making a set of assumptions about the real world sufficient for the purposes of characterising user-system interactions within a grammatical or logical notation. Although potentially testable, the initial derived principles may be idealised and relatively unconstrained by assumptions concerning non-ideal user behaviour. In contrast, a more empirical strategy might seek to constrain the initial formulation of principles on the basis of assumptions concerning the observed occurrence and distribution of particular categories of user behaviour with software products.

This chapter will illustrate a strategy in which the initial formulation of principles is constrained by emergent empirical phenomena. Laboratory experiments simulating key aspects of applications are one of the most important ways in which an acquisition representation is used for the purposes of gathering data. They enable us to establish and explore potential regularities in performance that can be attributed to factors present in the context of computer use. Naturally, experimental work makes many simplifying assumptions about the real-world of user-system tasks and contexts. Such work nevertheless provides a basis for clarifying assumptions, for an appropriate synthesis of theoretical structures within the science base, and for the subsequent generation of an effective applications representation.

We shall review three series of experiments. These will be described and discussed primarily in terms of empirical issues. Details of the synthesised theory and applications representation are elaborated elsewhere. The theoretical synthesis makes use of Interacting Cognitive Subsystems, a functional model of the representational and processing resources of human cognition (Barnard, 1985). From this model a methodology has been derived for analysing the operation of mental resources in the specific context of human-computer interaction (Barnard, 1987). This in turn has formed the basis of an explicit applications representation: approximate models characterising the way in which the human information processing mechanism will operate in the course of human-computer interactions. In order to achieve adequate scope, the idea is to generate a complete family of inter-related models reflecting the different levels of user expertise and different phases of cognitive activity that occur within extended tasks.

This theory-based approach to the construction of an applications representation has been realised in a demonstrator expert system (Barnard et al, 1986, 1987). Justified on the basis of regularities established through experimental research, its knowledge base represents principles derived from the science representation together with explicit assumptions to permit a mapping to the real world of interactive tasks. This form of applications representation thus offers the prospect of a decision aid for assessing how factors, within the overall context of computer use, contribute to productive and efficient user-system interactions.

4.1.2 Names as a case study for an applicable science

The three series of experiments to be discussed below examine the influence of command name sets on learning and performance in a text-editing task. Command name sets are a potentially productive arena for the experimental exploration of behavioural regularities in human-computer interaction. As a central feature of many forms of communicative dialogue, names and naming have been extensively examined in a number of core disciplines such as linguistics, philosophy and empirical psychology (e.g. see Carroll, 1978). In spite of such extensive academic interest, the potential applicability of our knowledge of naming to human-computer dialogues was in need of clarification. One thing was clear. Designers of early command languages developed forms of dialogue that caused considerable problems for users. Many of these problems could be attributed to the demands of learning, remembering distinguishing or otherwise accurately using system terminology. Both formal and informal observational analyses repeatedly identified names and naming as an important practical problem (e.g. Norman, 1981; Hammond et al, 1980; Long et al, 1983; Mack et al, 1983).

Given the users' difficulties with terminology and a body of knowledge from core disciplines, there is considerable scope for relating an existing science base to the real world of human-computer interaction. The earliest work attempted to do two things. First, several researchers tried to produce clear demonstrations that the way commands were named exerted an influence on user learning and performance (e.g. Ledgard et al, 1981). Second, researchers tried to isolate individual factors that might influence user behaviour. These included variables that have played an important part in basic research in cognitive psychology, such as word frequency and the dimension of concreteness/abstractness (e.g. Black & Moran, 1982),

Other variables included more linguistically oriented definitions such as the semantic specificity of terminology (Rosenberg, 1982) or the systematic use of linguistic relations among names within a command name set (e.g. see Carroll, 1982). Yet others focused upon more pragmatic variables such as the naturalness or computer-orientedness of command names (e.g. Landauer et al, 1983; Scapin, 1981).

In reviewing the literature and discussing its design implications, Barnard & Grudin (1988) point out that the success of these early studies was limited. The relationship between individual factors and performance proved far from straightforward. The issues not only concerned the interpretation of the main experimental comparisons and results, they also involved broader methodological considerations. So, for example, some studies gathered evidence using paper and pencil tasks to examine learning and memory for command names (e.g. Scapin, 1981; Black & Moran, 1982). Other studies used text-editors but with highly restricted functionality (e.g. Landauer et al, 1983). In addition, the studies reported in the literature used a variety of different techniques for deriving their experimental name sets, for assessing performance and for sampling both system functionality and user populations. Most studies have concentrated on text-editing, where the order in which commands are issued is determined by the corrections to be made in the to-be-edited text. Other studies have examined command names in tasks where users are required to learn strictly ordered sequences of operations (e.g. Barnard et al, 1984).

Barnard & Grudin (1988) provide a detailed discussion of the difficulties involved in interpreting the experimental literature on command names. In essence, the various research projects have used very different assumptions concerning tasks, definition of variables, performance measures and so on. In terms of Long's framework, they were creating different acquisition representations and using them differentially. Subsequent studies have attempted to understand the complex interactions of factors by examining more closely the contexts and tasks within which command use is embedded, the strategies that users adopt, and the cognitive consequences of employing them. The real challenge here is to establish how all the various factors inter-relate to determine user behaviour. Although naming represents only one part of real requirements for task and dialogue design, it constitutes a useful case study. It illustrates how a particular set of assumptions underlying an applied experimental paradigm can be utilised iteratively to build up a knowledge base of empirical regularities for interpretation within the science base.

4.2 The experimental paradigm

For the purposes of creating an acquisitions representation, prototypical text-editing in the real world can be assumed to include the following three basic stages. It involves the creation or retrieval of a document, the actual editing of the material, and the subsequent storage or output (printing/e-mail) of the revised document. Such activities occur within very different real world contexts (e.g. editing a scientific paper may involve a different procedure if the author does the text-editing than if a secretary corrects marked up errors). We sought to sample the learning of a text-editing system which embodied the basic stages but making use of only a restricted range of document processing tasks and system functionality. The objective of our text-editing system was to provide a simulation of the dialogue exchanges that are likely to occur within a command based editor and to provide appropriate means for assessing learning and performance.

The simulation was also designed to enable the control and assessment of particular factors in such a way that meaningful conclusions concerning their consequences could be derived. With more sophisticated systems involving full functionality this is often difficult if not impossible. For example, changing one aspect of dialogue style may have consequences for several other aspects in a way that cannot be controlled experimentally. Similarly, a full system may permit users to achieve (or fail to achieve) task objectives in several different ways. This potentially contributes sources of variability that may occlude the effects of the factors that are experimentally manipulated, although as a result the scope of those manipulations is restricted.

Experimental restrictions on text-editing conditions and tasks naturally limit the kinds of inferences that can be made concerning the real world of document processing. To the extent that it samples and captures key features of command name use in a fully interactive context, the adequacy of our acquisition representation will be assumed for the initial generation of input to the science base. Within the larger project of which this work is a part, various other forms of experimental simulation have been employed that bear on naming issues. These effectively extend the overall scope of our broader acquisition representation. They have examined the learning of lexical commands in different structural and stylistic forms of human-computer dialogue (e.g. see Barnard & Hammond, 1982; Barnard et al, 1984; Hammond et al, 1987; Barnard, MacLean & Wilson, 1988). The text-editing simulation described below was specifically created to examine the properties of command names and name sets in isolation from other structural or stylistic concerns of dialogue design.

To enable the experiments to be conducted on a reasonable time scale, the system, tasks and materials were also designed to enable novice users to tackle dialogue sequences without extensive training or documentation. Editing tasks were designed so that operations could readily be inferred by novice users. The material consisted of well-known single sentence proverbs such as "a stitch in time saves nine". These were initially presented to users in a distorted form, for example:

> sti tch in a time xpqy aves nine

Consistent with the three stages of prototypical editing, each basic task involved retrieving a distorted proverb, performing editing operations until it matched its well-known form and storing the result.

For this particular experimental paradigm, there were ten functions that operated on the proverbs. These were subdivided into five pairs of related operations. The first and last stages of prototypical text-editing were represented by a single pair of operations. These retrieved the proverb to-be-edited and stored the fully edited version. The second and main stage of actual editing was represented by four pairs of operations that modified the text of the proverb. These included the removal and addition of a word; the moving of a word from its current position to become either the initial word or final word in the proverb; the addition of a letter to the beginning or end of an individual word; joining two separate word segments into a single segment or splitting a single segment into two. Name set manipulations were carried out over these ten functions.

In addition to the ten functions that operated on the proverbs, there were two additional functions whose names remained constant across conditions and experiments. These were provided to enable users to obtain assistance or to rapidly correct errors. A "help" function was available to provide information on the command names and their operations, while a "cancel" function enabled them to "undo" their last action whether completed or not. Both of these functions also formed a basis for dependent measurement.

Editing involved a display with four main areas. These are shown in Figure 4.1. The uppermost area was used to display a menu of the available commands. The second area contained information about the target proverb. At the beginning of a trial, this area displayed a target number, information that had to be given in order to retrieve a proverb. Once retrieved, a target proverb state was continuously displayed until the proverb had been successfully edited. At this point, information would be displayed indicating that the target state had been achieved. Carrying out the storage operation which terminated a trial would then cause this panel to display the target number of the next proverb to be retrieved and edited.

The third area initially displayed the unedited proverb. It also recorded the history of editing operations. As each operation was performed, the new versions would be successively listed below the old ones. When this area was full, the oldest scrolled out of sight. Users could return to any prior version by invoking the cancel operation.

The fourth and bottom area on the screen was the command entry area. This contained action prompts and echoed user keystrokes. In order to carry out a function, users typed in the name and terminated this entry with a carriage return. The system would then prompt users as to any arguments required by that command which would then be entered in a similar fashion. In Figure 4.1 the command "join" is being entered. During argument entry this field contained a reminder as to which command was being used. Error messages would also appear here (e.g. 'unrecognised command').

In the basic editing state, the uppermost panel shown in Figure 4.1 contained a reminder concerning the help and cancel functions. In order to obtain a command menu, such as that shown in the figure, users would have to invoke help. A second help request from this menu state caused the full editing screen to be completely replaced by a new full screen display in which definitions of command actions were listed beside the individual

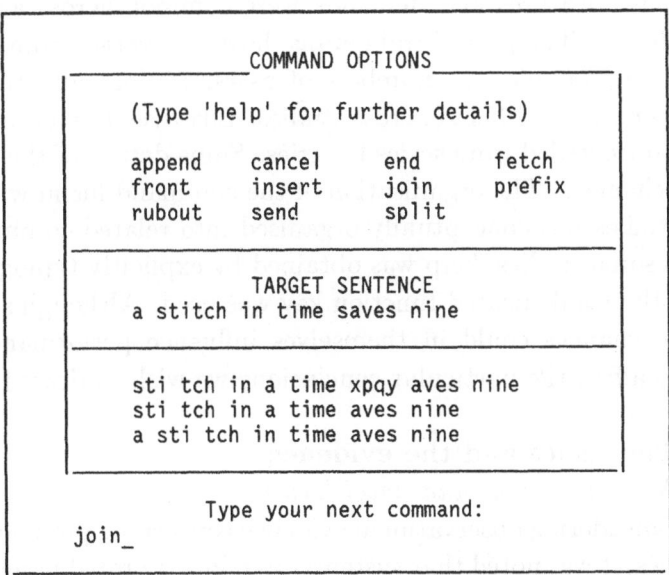

Figure 4.1: A display state from the laboratory text-editor illustrating the four main areas, the uppermost showing the menu for a set of specific command names (from Barnard et al, 1982). Copyright 1982 Taylor & Francis Ltd, and reproduced by permission

command names. Commands could be entered either in the basic editing state or with the menu of command options. They could not be entered when the full listing of command definitions was displayed.

On any given trial, a proverb could be successfully edited by issuing six commands. These consisted of retrieving the proverb, making four changes, and storing the final version. The set of proverbs used as experimental material was designed such that the optimal frequency of using the eight editing operations was equated across trials. Suboptimal command sequences were nevertheless possible. Users could, for example, "move" material by "deleting" it and "inserting" it elsewhere. The system preserved a time stamped session log of all user input, from which a variety of measures were abstracted, including errors, viewing times, command frequencies, as well as the use of the help and cancel facilities.

From this overall description of the laboratory editing system, we can now summarise the basic cognitive demands involved in learning to use it. In addition to learning the command names and operations, users had to learn about the basic structure of the task, i.e. retrieve-edit-store, about properties of the system such as the help facilities and the effects of the cancel operation. They also had to infer which of the eight editing operations were required for a given proverb and to know how to interpret any system prompts concerning arguments to be entered or error messages.

These basic features and demands were common across all the studies reviewed below. The procedural details, however, varied from one study to another. For example, the numbers of users, sessions and the number of proverbs per session differed across studies. Likewise, the initial familiarisation procedure varied from series to series. Some details of the system itself were also changed. The organisation of the command menu was alphabetic in some studies but conceptually organised into related command pairs in others. In some studies, help was obtained by explicitly typing a command word, in others a dedicated function key was used. Although some of these procedural changes could in themselves influence performance, they are unlikely to affect the particular conclusions we wish to draw here.

4.3   The issues and the evidence
4.3.1   Novel meanings and novel forms

Considering observational evidence concerning problems with command names, it was noted that systems often incorporated names with novel meaning or novel form. In the case of command meanings, use of jargon terms abounded. In some cases, the use of command names would involve making use of a feature of a word in our normal vocabulary to signify a relevant feature of the command operation (e.g. KILL). In some instances (e.g. CAT), a command string could have a meaning in everyday language totally unrelated to its sense or referent in the computing context (e.g. see Grudin & Barnard, 1984a).

In the case of the form of command names, abbreviations were commonly used. These were motivated at least in part to reduce the number of keystrokes required to enter a command. In seeking a set of abbreviations to apply over a large range of operations, it was often difficult to maintain either a straightforward abbreviation rule or a clear mnemonic relation between a particular operation and the command abbreviation. In certain instances, where some mnemonic relation may have motivated an abbreviation, it was far from clear that users would be able to infer and make use of its derivation.

Indeed, studies contributing to the science base associated with human learning and cognitive psychology suggested that performance should be influenced by the precise demands for learning novel form or meaning. For example, Landauer et al, (1983) pointed out that command names with some natural association to their stimuli and with well-known response patterns should be easier to learn. Based on such reasoning, they carried out a study within the more applied setting of human-computer interaction, and failed to show the clear pattern that might have been expected from the science base. In their study, the effects of different techniques for generating names (user nomination, designer choice, or random name assignment) were assessed during the initial learning of a text editor. The precise way in which the name sets were generated did not lead to large or statistically reliable differences. In this kind of natural learning setting, there are many difficulties that naive users must overcome during their initial encounters with a text editor. Name choice, although it might contribute positively, did not appear to help a great deal.

In the Landauer et al study, and in others which addressed similar issues (e.g. Ledgard et al, 1981; Hammond et al, 1987), users learned command dialogues in which command names were included within a command argument syntax. Indeed, there were several clues that the learning of names and the learning of command structures were not independent of each other (e.g. see Landauer et al, 1983; Hammond et al, 1987). As noted earlier, the proverb editing task provides a context where users do not need to generate command names within a syntactic structure. The first study we shall review (Grudin & Barnard, 1984a) sought to examine systematically how novel form and novel meaning might influence learning and performance.

In this study, Grudin & Barnard examined five different types of vocabulary for text-editing (Figure 4.2). One of these vocabularies, specific names, involved known words whose meaning outside the computing context was related to the effects of their underlying operations. The differences between the effects of the opposing operations were marked by using congruent terms (cf. Carroll, 1982) in natural language (e.g. front/back). The remaining four sets were designed to examine the cumulative effects of having to learn novel form and meaning.

With the vocabulary of abbreviations, users had to learn a novel form. Here, a contraction rule was applied and the full command names were reduced to the first three different consonants in the full name (hence "pnd" rather than "ppn" for append). This particular rule permitted a direct contrast with a vocabulary of consonant strings (see below). With the vocabulary of unrelated names, users were required to learn novel meanings. Since the words could be expected to be in a users' natural vocabulary, they were not required to learn a novel form. Rather, they were required to associate that form with a completely novel meaning in the computing context.

As with the unrelated names, those asked to learn pseudowords had to acquire new meanings, but they also had to acquire novel lexical forms. As with the abbreviations, users of the consonant string vocabulary had to acquire a set of three letter response patterns, but this time in the absence of any prior lexical or meaningful associations.

In terms of overall measures of learning and performance, the results of this experiment were clear. The vocabulary of specific command names gave rise to the fastest times (see Figure 4.3), the lowest number of transactions and least frequent use of the help facilities. There were statistically reliable costs, of the same general magnitude, associated with both novel

Specific Names	Abbreviations	Unrelated Names	Pseudowords	Consonant Strings
front	frn	stack	blark	ksd
back	bck	turn	lins	rip
insert	nsr	afford	aspalm	trp
delete	dlt	parole	ragole	fnm
prefix	prf	chisel	clamen	drs
append	pnd	uncurl	extich	gpf
merge	mrg	rinse	dapse	lts
split	spl	throw	thrag	ctf
fetch	ftc	light	manst	nrb
store	str	brake	fluve	rcn

Figure 4.2: The five command name sets from Grudin & Barnard (1984a) Copyright 1984 by the Human Factors Society Inc, and reproduced by permission

forms (abbreviations) and novel meanings (unrelated names). Even greater costs in learning and performance were evident when both novel meaning and novel form needed to be acquired (pseudowords and consonant strings).

Furthermore, detailed breakdown of individual performance measures revealed patterns that appeared to reflect the precise demands incurred. So, for example, the non-lexical command forms (abbreviations and consonant strings) gave rise to high frequencies of command entries that the system could not recognise. Lexical command forms (specific, unrelated and pseudowords) gave rise to lower frequencies of unrecognised commands. Users of the non-lexical forms also had the menu of command names displayed for a greater percentage of the total editing time than did users of the lexical forms. Likewise, those command vocabularies in which novel meanings had to be acquired gave rise to the highest proportion of instances where users consulted the menu panel of command names and then went on to examine the full help panel listing names with definitions of their associated operations.

Other findings indicated that pre-existing knowledge of lexical form assisted learning and memory. Thus, performance with the pseudoword vocabulary began not unlike that for consonant strings, but by the end of the third session of editing more closely resembled the level and nature of performance attained with the unrelated command vocabulary. By this stage, it seems that the pseudowords were effectively lexicalised. In contrast, performance with the consonant strings remained poor throughout. Indeed, on a paper and pencil recall test taken during each session, users of the conso-

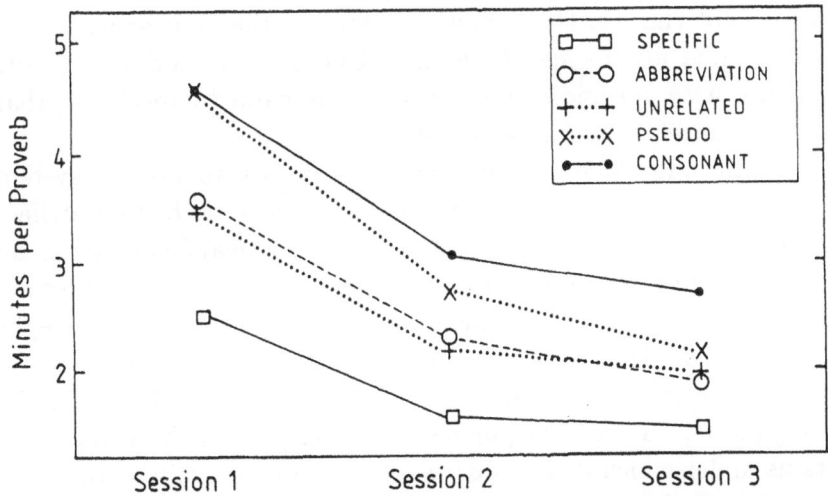

Figure 4.3: Mean times taken to complete the processing of each proverb for the five command name sets over three editing sessions (from Grudin & Barnard, 1984a). Copyright 1984 by the Human Factors Society Inc, and reproduced by permission

nant strings averaged only 34% correct name recall. Recall for the abbreviations, pseudowords and unrelated names averaged between 79% and 86%, while that for the specific names average 97% correct. In contrast, recall for operations was near ceiling (94%), indicating that there were real differences in the ease of remembering the different types of names required for the same operations.

Landauer & Galotti (1984) pointed out that the particular pattern of results could have been influenced by the way in which the name sets had been selected and by the design of the help system. The help panel which listed the command names and their definitions used the same definitions for all five sets. These definitions were phrased in terms of the specific command names. This ensured that users of the abbreviations would know the full name. A consequence of this strategy was that the command names were repeated in this panel for the specific name set but not for the other four name sets. The repetition could have given the users of the specific name set an advantage in learning name-operation mappings inattributable to the actual type of name involved.

Were this to prove the case, the experiment might simply have shown that command names involving a known lexical form (specific, abbreviations, and unrelated words) were better than nonsense (pseudowords and consonant strings). For the purposes of building principles within the science base, an ambiguity of this type is unsatisfactory. Thus, in order to clarify the point, two further groups were tested in a single session experiment (unpublished). These groups learned the specific command name vocabulary and the unrelated name vocabulary. Exactly the same procedure was followed except that, in the definitions listed in the help panel, synonyms were substituted for the specific names. Under these modified conditions, performance with specific command names remained superior to that for semantically unrelated command names.

The Grudin and Barnard study gave rise to an apparently systematic pattern of data. On the assumption that there were no further artifacts of the sort examined in the follow up study, the empirical facts seem in accordance with the more general claim that poorer performance will result as a function of increasing the demands to learn novel forms or novel meanings. Thus, the general claim constitutes a candidate empirical regularity for incorporation and interpretation within the science base. The incorporation of the regularity, naturally requires supporting detail referencing both the conditions under which it was observed and the finer grain features of performance associated with the specific demands involved.

As a potential component of the science base a candidate empirical regularity directly relates a descriptive feature of dialogue elements to user learning and performance. This kind of inter-relationship is itself open to further refinement, empirical evaluation and generalisation. Its generality could, for example, be tested by examining the learning of other kinds of novel symbolic forms such as visual icons. However, here we shall distinguish a candidate empirical regularity from a candidate scientific principle. This offers a potential explanatory account of the observed regularity, contributes directly to the enrichment of the science base, and functions to provide a firmer basis for prediction in novel circumstances.

The particular empirical regularity that emerges from the Grudin and Barnard experiment has been interpreted as arising out of underlying requirements to co-ordinate and control the activity of separable processes within the human information processing mechanism at different stages of learning (Barnard, 1987).

Barnard argues that the entry of computer commands is co-ordinated and controlled within a predefined configuration of mental processes. A configuration is not unlike classic 'stages' of processing. Included within the configuration for command entry are one mental process which generates form from meaning and another which uses that form to generate a representation to control keystroking. These processes can, however, only operate automatically if they contain well established procedural knowledge for carrying out the mappings required for a particular command set. So, for example, the lexical form "delete" can be automatically generated from a representation of the meaning of the required operation such as "excise an entity from a piece of text". In the absence of well-established procedures, the cognitive mechanism may have to access declarative memory structures concerning prior usage or utilise other inferential processes to determine a possible output mapping. The poorer learning performance is assumed to reflect these additional cognitive activities and their demands for additional information. Thus, the candidate scientific principle involves assuming a direct relationship between the extent of additional cognitive activity and the number of mental processes for which new procedures need to be established.

At one extreme, specific command names require little in the way of acquiring new procedures for mapping form to meaning. At the other extreme, pseudowords and consonant strings require the development of new procedures in both the processes described. Nevertheless, learning progressed more rapidly with pseudowords than with consonant strings indicating that the candidate principle just described would require supplementing with a second principle concerning the acquisition of new mappings within individual mental processes. In this case, the relative advantage for pseudowords can be understood by assuming that known structural descriptions (lexical form) can assist in the recovery of information from memory during the acquisition of new mappings (see Barnard, 1987).

The two principles concerning memory access and the number of processes for which new procedural knowledge is required are obviously only fragments of what needs to be present in the science base to permit wider generalisation and prediction. The next section will examine how similar empirical regularities can be used to elaborate constraints on the acquisition of new procedures within a single stage of mental processing, the generation of form from meaning.

4.3.2   Contextualisation of meaning and set effects

The kinds of name sets examined by Grudin & Barnard deliberately involved gross distinctions. Where users were required to learn new meanings (unrelated names, pseudowords, consonant strings) any of the command names could have been assigned to any of the operations. Such complete name-operation ambiguity, particularly for word forms, is rare. More typically, users have some semantic knowledge which may pragmatically restrict name assignment. A separate series of studies focused upon command names whose meanings bore some systematic relation to the operations they invoked in the computing context. In these studies, the nature of the semantic relationship was examined as was the way in which individual command names were organised within sets. For these studies, the emphasis was therefore upon contextualising the meaning of known words for the computer editing task. Since known words are involved, the interest for the science base lies in understanding how meaning is used to generate known words in this kind of applied setting.

The first of these studies (Barnard et al, 1982) compared semantically specific command names, almost identical to those used by Grudin & Barnard (1984a), with a set of semantically more general command names. The two vocabularies compared are shown in Figure 4.4. The semantics underlying general words like "change" or "move" cover a whole class of potential operations. More specific terms such as "substitute" or "advance" invoke additional semantic information concerning the nature of the operation. Thus, "substitute" presupposes "change" and also conveys the additional information "replace with an entity of a similar type." Likewise, "advance" presupposes movement with the addition of a directional feature. The presence of additional semantic features might function to enhance learning and memory. An independent paper and pencil study confirmed that the specific names were judged to be more appropriate for their associated operations than were the general terms.

The choice between general and specific commands raises real uncertainties. Arguments can be mustered in favour of performance advantages for both types of command name. An interesting practical discussion of the utility of generic commands is provided by Rosenberg and Moran (1985). The science base within cognitive psychology also gives rise to potentially opposing arguments. For example, cued recall for sentences containing specific verbs has been shown to be superior to cued recall for sentences containing general verbs (Thios, 1975). Similar effects have been shown for general and specific nouns (Gumenik, 1979). On the other hand, general terms can be used in a broad range of contexts without violating their natural implications. They also tend to be of higher frequency in the language and hence easier to generate from memory than low frequency terms. Since their meaning in context also has to be actively assigned, this might require more mental processing activity than the specific words and hence to more effective memory traces (cf. Craik & Lockhart, 1972). Weighed against these possibilities is the knowledge that in long term retrieval tasks people are likely to confuse words with similar meaning (Baddeley, 1966). In order to eliminate the frequency factor, the general and specific terms shown in Figure 4.4 were selected so that the samples did not differ on standard measures of word frequency.

The experiment was designed to clarify the potential role of semantic specificity in the context of text-editing. Two groups of users learned the specific command name set and two learned the general command name set. For each command name set, one group (supergoal conditions) were shown the fully corrected proverb as the target state (as in the second panel

Specific	General
fetch	transfer
send	put
front	move
end	shift
insert	add
rubout	edit
prefix	affix
append	attach
join	restore
split	open

Figure 4.4: The sets of specific and general command names studied by Barnard et al, (1982). Copyright 1982 Taylor & Francis Ltd, and reproduced by permission.

down of Figure 4.1). The other groups were constrained to carrying out the editing operations in a particular sequence (subgoal conditions). Thus, the target state specified could be achieved by a single operation. Once carried out, the target state would change to indicate the next operation to be performed. This manipulation was included because previous research had produced some evidence that a requirement to plan operations, or to learn sequences, might influence the way in which name learning occurred (e.g. Hammond et al, 1980; 1987).

The experiment consisted of two sessions. In the first, users were given some initial verbal instruction on the system and its use. They then edited a sequence of proverbs. About a week later they returned for the second session in which they first carried out a memory test, like that employed by Grudin & Barnard. They then completed two individual difference tests, the cognitive failures questionnaire (Broadbent et al, 1982) and a specially constructed test designed to assess how readily individual users could generate lexical items from a representation of their meaning. In this test they were shown definitions of nouns and verbs which varied in their concreteness and frequency. The users' task was to generate the word to which each definition referred. The users completed the session by interactively editing a further sequence of four proverbs. For these proverbs, all groups carried out the editing with the final target proverb shown. Thus, users in the subgoal condition for the first session, worked under supergoal conditions in the second one. They were effectively required to transfer their initial training to new conditions in which they had to plan their own sequence of editing operations.

The effects of giving the users subgoal or supergoal targets did give rise to systematic effects on performance. This factor did not influence the overall frequency with which the help system was consulted. However, once the menu of command name options was consulted, then users in the subgoal condition were more likely to enter a command from this context than were users in the supergoal conditions. Users in the subgoal condition were also more likely to cancel the effects of a command operation. This suggested that a more impulsive "act-and-cancel-if-inappropriate" strategy were more readily established in the subgoal, rather than the supergoal condition. However, these effects in no way interacted with the effects of command name type.

Two of the more revealing performance measures are reproduced in Figure 4.5, the mean times taken to solve each proverb and mean frequency of use for the second level help (names plus their associated operations). In Session 1 users of the general command set took longer on average to edit each proverb (243.5 s) than did the users of the specific name vocabulary

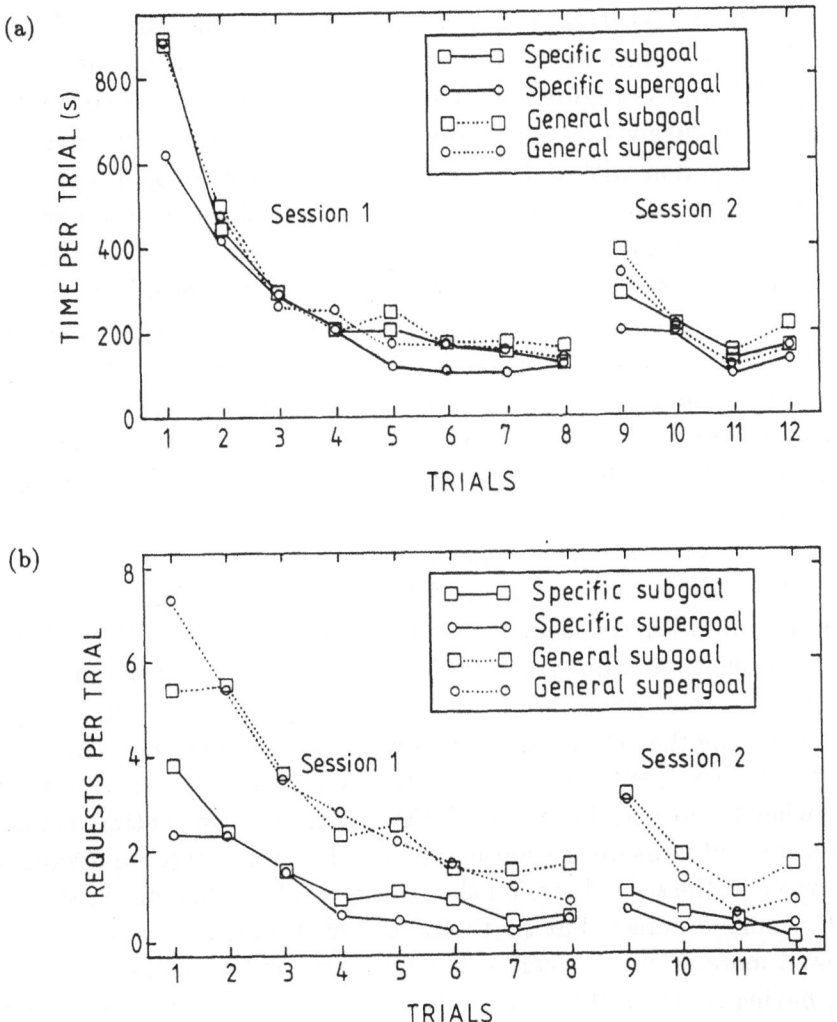

Figure 4.5: Data from Barnard et al, (1982) for specific and general command sets under the two 'goal' conditions. These show (a) mean time taken to edit proverbs within each session and (b) Mean frequency with which users requested the help panel showing the command name set with definitions of their associated operations. Copyright 1982 Taylor & Francis Ltd, and reproduced by permission

(209.5 s). This difference was not statistically reliable. For Session 2 users of the general command names took significantly longer, but this effect interacted with the particular trial. The temporal disadvantage for the general command names in session two was confined to the first trial of that session.

Although the overall time taken was not substantially affected by the type of command name, the ways in which users went about the editing task were affected. As Figure 4.5b shows, users of the general vocabulary were much more likely to request the help panel showing the command names in conjunction with definitions of their associated operations. At the beginning of each session, this was a substantial effect. It did, however, diminish reliably within each session. More detailed analysis of the patterns of performance suggested that users of the general command vocabulary not only needed more assistance, but also developed a strategy to obtain it quickly. Under circumstances where no help was required, the time taken to enter a command was equivalent for both command sets. However, when help was consulted, users of the general command set issued the help command with shorter delays than did the users of the specific command name vocabulary.

The semantic cues in the specific command names were obviously of value, in the sense that users required less frequent assistance and spent less time overall with the help panels displayed. However, the value of semantic specificity was not translated into greater speed of performance. Although users of the general command name set spent proportionally more of their time using the help facilities, this was offset by the time users of the specific vocabulary spent prior to consulting help. Indeed, active consideration of what to do may have helped these users to form better memory representations of name-operation mappings. The recall test conducted at the beginning of the second session showed no reliable difference in the free recall of command names, but that users of the specific command vocabulary showed more accurate recall of the actual editing operations.

Thus, having semantically specific names for pairs of related operations reduced the users' requirement to access the help facilities and enhanced memory for the operations after a week's interval. For the purposes of establishing empirical regularities and interpreting them within the science base, it is important to establish whether semantic specificity per se is the key factor. The strategic aspects of performance suggested that users of the general command names were well aware that they did not know which name to use for a particular operation and relied on the relevant help facility. In learning the general or specific command names, the users were not simply contextualising the meaning of known words for particular command operations. They were also learning them as part of a wider set of name-operation mappings. Under these circumstances, the specific set is associated with less overall ambiguity in name-operation mapping than the set of more general command names.

For the latter set several of the names assigned within a pair were such that they could readily be exchanged (e.g. move/shift; affix/attach) – or even between pairs (affix-add; transfer-move etc.). With the semantically specific vocabulary, there was much less potential ambiguity in name-operation mapping. So, for example, although the two pairs front/end and prefix/append were potentially confusable, the set appeared to function well to discriminate both the types of operation performed by each pair and to discriminate within pairs. Thus, semantic specificity could be functioning indirectly by affecting the overall level of ambiguity in name-operation mapping.

MIXED WITHIN PAIRS		MIXED BETWEEN PAIRS	
CMD BAL 1	CMD BAL 2	CMD BAL 1	CMD BAL 2
Front	Move	Front	Move
Shift	End	End	Shift
Insert	Add	Insert	Add
Edit	Rubout	Rubout	Edit
Prefix	Affix	Affix	Prefix
Attach	Append	Attach	Append
Join	Restore	Restore	Join
Open	Split	Open	Split
Fetch	Transfer	Fetch	Transfer
Put	Send	Put	Send

Figure 4.6: Two representative counterbalancings for each of the two types of mixed command sets studied by MacLean et al, (1988)

A follow-up study (MacLean et al, in press) provides some clarification of this issue. It examined the learning of name sets that were made up of half general and half specific names. In one condition, the general and specific names were systematically mixed within pairs. Thus, for each pair of operations one command was assigned a specific name and the other a general name. In a second condition, two pairs of operations were assigned specific names and two pairs were assigned general names. The remaining pair (always the filing operations), was assigned one general and one specific name. The various combinatorial possibilities were covered by eight counterbalanced versions. Two representative versions of each kind of mixed set are shown in Figure 4.6. The text-editing procedure was exactly the same as that used in the "supergoal" conditions of the earlier experiment.

If semantic specificity were functioning to assist the learning of name operation mappings, then both forms of mixed name set with only fifty percent specific names should be inferior to the wholly specific set studied earlier. Furthermore, it would be expected that users would consult the help facilities more frequently prior to entering a general command than prior to entering a specific command. Any effects of this sort might nevertheless be influenced by the wider constitution of the set. Where general and specific names are mixed within a pair of operations, knowledge of the specific member of the pair should enable the users to infer what effect the general command would have. This in turn should reduce their requirements to proceed to the second help panel which linked names to the definitions of the operations. Where the name set includes pairs of related general names, the requirement for help should persist relative to the pairs of specific names.

Figure 4.7a shows the average editing times for these two types of name set and contrasts them with the times taken for the wholly general and wholly specific sets of the previous experiment. Figure 4.7b shows the same contrasts for the use of the help panel in which command names were listed together with the definitions of the operations they performed. As far as the overall levels of performance are concerned, the pattern is clear. The two forms of command name set containing five general and five specific names gave rise to equivalent levels of performance throughout learning. Furthermore, that level of performance closely approximates that for the command name set made up wholly of specific names.

In addition, on key measures of accuracy and help usage there was little evidence of appreciable differences between general and specific names within the sets. Where pairs of related operations were assigned one general and one specific name, users were apparently able to infer the name-operation mappings. Presumably, users could utilise the specific name to derive the effect of its more general companion without recourse to the help system. Where general names were used to refer to only two pairs of related operations, users were also able to cope adequately – most plausibly by explicitly encoding the name-operation mappings in their own memory representations.

Taken together with the patterns of data exhibited in the two previous experiments, these results would suggest that semantic specificity, in and of itself, is rather less important than the effective degree of ambiguity in name-operation mapping within the complete set. In the Grudin & Barnard study, learning names whose meaning was totally unrelated to their effect in

(a)

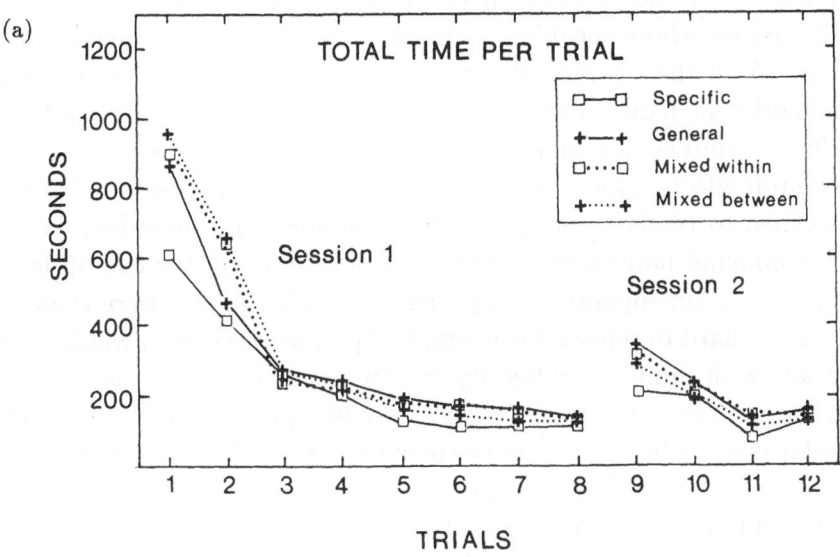

(b)

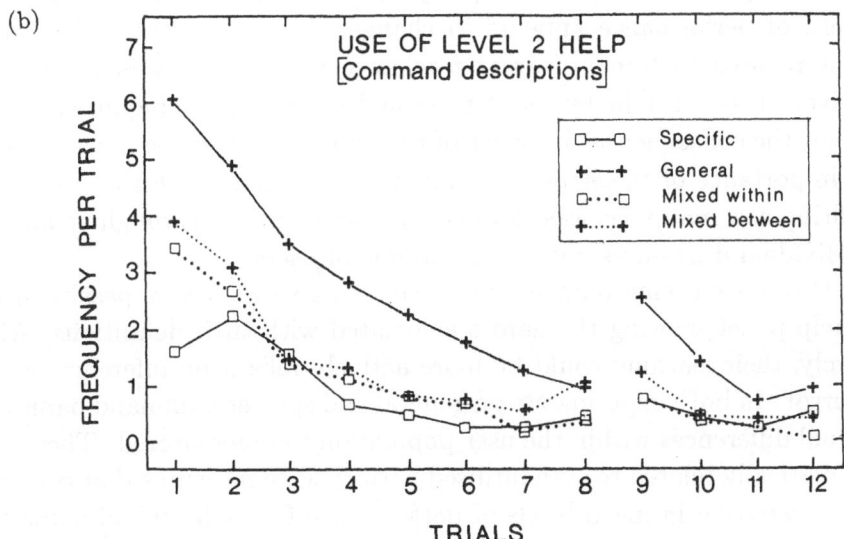

Figure 4.7: Performance with the mixed command name sets (MacLean et al, 1988) compared with data from the Barnard et al, (1982) study, showing (a) total time per trial and (b) use of the help panel with command names and their associated definitions (HELP2)

the context of text editing involved maximum ambiguity of name-operation mapping. This had negative performance costs relative to semantically specific names whose meaning was clearly related to the operations they invoked. With the semantically general command terms, the overall level of ambiguity in name-operation mapping was roughly mid-way between that for the unrelated command name set and the specific name set. The behavioural effects appeared to be primarily strategic. Users relied on the help system to resolve ambiguities in name operation mapping. With the mixed command name sets studied by MacLean et al, the overall levels of ambiguity in name-operation mapping, although present, were even lower. Here, it was hard to separate learning and performance from wholly specific name sets with minimal ambiguity in name-operation mapping.

On the assumption that these effects would replicate with other materials in similar task contexts, the candidate empirical regularity for incorporation in the science base is best expressed in terms of the overall level of name-operation ambiguity. Given the evidence available it would be inappropriate to use precise metrics. In outline form, minimal levels of ambiguity appear to be readily resolvable. As ambiguity becomes significant, so the strategic pattern of performance appears to change. In the current task context, this is realised in terms of greater reliance upon help systems, but with relatively little cost in terms of time and error. As ambiguity increases further, the consequences in terms of time and error become more extreme. The importance of the strategic component in coping with the problem of acquiring new mappings was clarified somewhat further through an analysis of individual differences within the sample of users.

In this system environment, users could make relatively passive use of the help panel showing the names associated with their definitions. Alternatively, their learning could be more actively reliant on inference or trial and error. In both experiments on general and specific command names, individual differences within the user population were monitored. These were systematically related to standardised performance measures that corrected for the variation in mean levels of performance for each type of name set.

Those who scored highly on the cognitive failures questionnaire (Broadbent et al, 1982) tended to make use of the full help panel less frequently than those with a low cognitive failure score. Similarly, those who were good at the definitions test also made less use of this help facility. These analyses are reported in Barnard et al, (1982). Further analyses (MacLean et al, in press) found these correlations in both experiments. However, a sample of the users were subsequently brought back to the laboratory to complete a form of IQ test. When IQ was partialled out, the definitions test added nothing. Hence, it appeared that the more astute users made less use of the full help facility than the less astute ones. Possibly the astute users were making use of more active learning strategies.

The correlation with cognitive failures scores remained significant even with IQ partialled out. In this case, it is possible that those users "who tend to go from one room to another and fail to remember why they went there" may have made little use of the help facility for different reasons. These users may have been prone to more impulsive action or otherwise not have bothered to capitalise on relevant information within the learning environment (see Barnard et al, 1982). Either way, the demands imposed by the differently constituted name sets may have functioned to modulate users predispositions to adopt different strategies for learning name-operation mappings.

In section 3.1, two candidate scientific principles were offered to interpret the main empirical regularities obtained by Grudin and Barnard (1984a). One of those principles related the extent of cognitive activity to the number of different mental processes for which new procedural knowledge needed to be acquired. Since both general and specific command names were known lexical items, within Barnard's (1987) analytic framework, users of these command vocabularies are faced with a demand to acquire new procedural mappings within just one process, that which generates lexical items (the command word) from a representation of the meaning of the operation to be performed. In the earliest stages of learning this should not be taken to mean that other representations and processes aren't involved. Indeed during the earliest phases of learning users are quite likely to call upon both inferential processes and relatively superficial declarative memory representations.

Given a representation of an operation, and a set of known names, a user's problem is essentially one of deciding which is the 'correct' mapping. This form of item uncertainty can be resolved in a number of ways. Naive users may draw an inference from their prior semantic or pragmatic knowledge, they may access a memory record of the introductory information or even a record of a prior error sequence. They may acknowledge the uncertainty and decide to consult the help facilities to the point where they have sufficient information to act, or they may decide to resolve the uncertainty by trial and error.

In the proverb editing paradigm all of these strategies probably come into play. Including a consideration of variation within the user population, the evidence appears consistent with the following more general candidate scientific principles: (1) where there is an unautomated mapping from meaning to form, the wider cognitive system configures its resources to resolve the uncertainty either by inference, direct memory access or by referring to external sources of information, (2) requirements for inferential processing and/or memory access increase with the overall extent of the uncertainty; (3) inferential processes evaluate strategic trade-offs between internal or external means of resolving uncertainty within individually determined limits.

Thus, where the overall ambiguity of name-operation mapping is minimal or small (specific and mixed command name sets), the strategic balance by and large favours internal resolution of item uncertainty. As item uncertainty increases (e.g. General names) the strategic balance moves more towards external sources (e.g. Help). With relatively high levels of ambiguity in name-operation, even external sources of information require considerable use of inferential processes and memory access. Under such circumstances very real performance costs start to emerge (e.g. unrelated names).

4.3.3 *Contextualisation of learning form*

Section 3.1 considered candidate empirical regularities and scientific principles governing the acquisition of novel forms and novel meanings. In section 3.2, we examined the acquisition of novel mappings between known meanings and known forms. In discussing the Grudin & Barnard (1984a) study, it was argued that cognitive demands were incurred not only in the acquisition of new name-operation mappings but also in the acquisition of novel command forms – such as abbreviations, pseudowords and consonant strings. When performance with their set of abbreviations was discussed, it was assumed that users could effectively map the full names onto their underlying operations. The demand, as such, was to learn the abbreviation rule. The present section will examine these demands in more detail with a view to formulating principles for the acquisition of mappings to novel form.

Abbreviated command names have been widely explored and again there is relevant input from basic psychology (e.g. see Hodge & Pennington, 1973). In the specific context of learning computer commands the emphasis has been upon the nature of abbreviation rules (e.g. see Ehrenreich & Porcu, 1982; Hirsh-Pasek, Nudelman & Schneider, 1982; Streeter, Ackroff & Taylor, 1983). Having a structurally consistent abbreviation scheme appears to be an important factor in determining user performance. However, abbreviation schemes are learned, or otherwise created, in a specific task context. That context is in part determined by the users' experience and in part by the actual function of the scheme. When examining a list of abbreviated file names it may be most helpful to have a scheme that supports the reconstruction of the full name. When abbreviating command names, it may be more helpful to have a scheme that supports memory for the abbreviation itself. These issues were examined in a further set of experiments using the text editing paradigm (Grudin & Barnard, 1984b; 1985). The studies were based around specific command names in order to minimise ambiguity in name-operation mapping attributable to potential confusions among the full names for the set of operations.

The main study (Grudin & Barnard, 1985) was motivated by a combination of theoretical and practical concerns. It compared the learning of an imposed, but structurally consistent, set of abbreviations with abbreviations that users generated for themselves. The interest in user created abbreviations arose from practical arguments concerning the desirability of permitting users to create their own command names. However, in addition to the contrast between abbreviation types, the way in which the abbreviations were introduced was also varied. Informal observation suggested that some users know an abbreviated command and its effects without apparently knowing the exact name from which the abbreviation was derived. Accordingly, it was argued that users might have a better grasp of name-operation mappings if they had initial experience with the full command names prior to transferring to an abbreviation scheme.

This experiment closely followed the system design and technique used by Grudin & Barnard (1984a). The principal procedural difference was that two editing sessions were carried out on the same day with a short break between them. The third session occurred approximately one week later.

Full Name	Predefined Abbreviation
front	fr
back	ba
insert	in
delete	de
prefix	pr
append	ap
merge	me
divide	di
fetch	fe
store	st

Figure 4.8: The set of specific command names used by Grudin and Barnard (1985), together with the two letter predefined abbreviations (truncation rule). Copyright 1985, by the Association of Computing Machinery Inc, and reproduced by permission

One group learned abbreviations of a semantically specific command name set from the start of the experiment. The abbreviations were all two-letter forms based on a simple truncation rule. They are shown in Figure 4.8 in conjunction with the full command names. The abbreviations were introduced during the initial computer assisted instruction which also referred to the full command names several times. This group was required to use these abbreviations. A second group was told prior to carrying out any editing that they could use either the full command names or the predefined abbreviations.

A third group were given different initial experience. They learned and used the full command names during the first session of editing but were transferred to the optional use of the predefined abbreviations for the remaining two sessions of editing. The final group learned and used the full command names for the first session. At the end of this session, these users were asked to create their own abbreviations and use them for the remaining sessions. These abbreviations, once created, were automatically incorporated into the help systems available over subsequent trials.

With the exception of one person, those users given the option of entering full names or abbreviations rarely used the full command names. Omitting this person, this group did not appear distinguishable from the abbreviation-only group and hence, for most analyses, the data for these two groups were combined.

The time and error data are shown in Figures 9a and b. In sessions one and three, the various groups did not differ in their editing times. It should be noted here that the two letter truncations gave rise to editing times that were equivalent to those for the group using the full command names. This contrasts with the difference found in the earlier Grudin & Barnard (1984a) study where the abbreviations were governed by rules for a more complex three letter contraction. In session 1, both forms of abbreviation scheme were, however associated with more command entry errors relative to use of the full specific command names (Figure 4.9b).

Those users who transferred from the full names to the predefined abbreviations did so without any detectable time cost at the start of session two. However, those users who created their own abbreviation scheme initially took considerably longer over the editing at the beginning of session two than did those groups using the predefined truncation rule. In this session users who created their own abbreviations made more use of the help facility and made more errors. The errors also persisted into session three. Indeed, across these sessions, users who created their own abbreviations made five times as many errors as the group who switched to the set of predefined abbreviations. Both groups started off with the full names and showed exactly comparable error rates during the initial session.

The real problem for users who created their own abbreviations was that

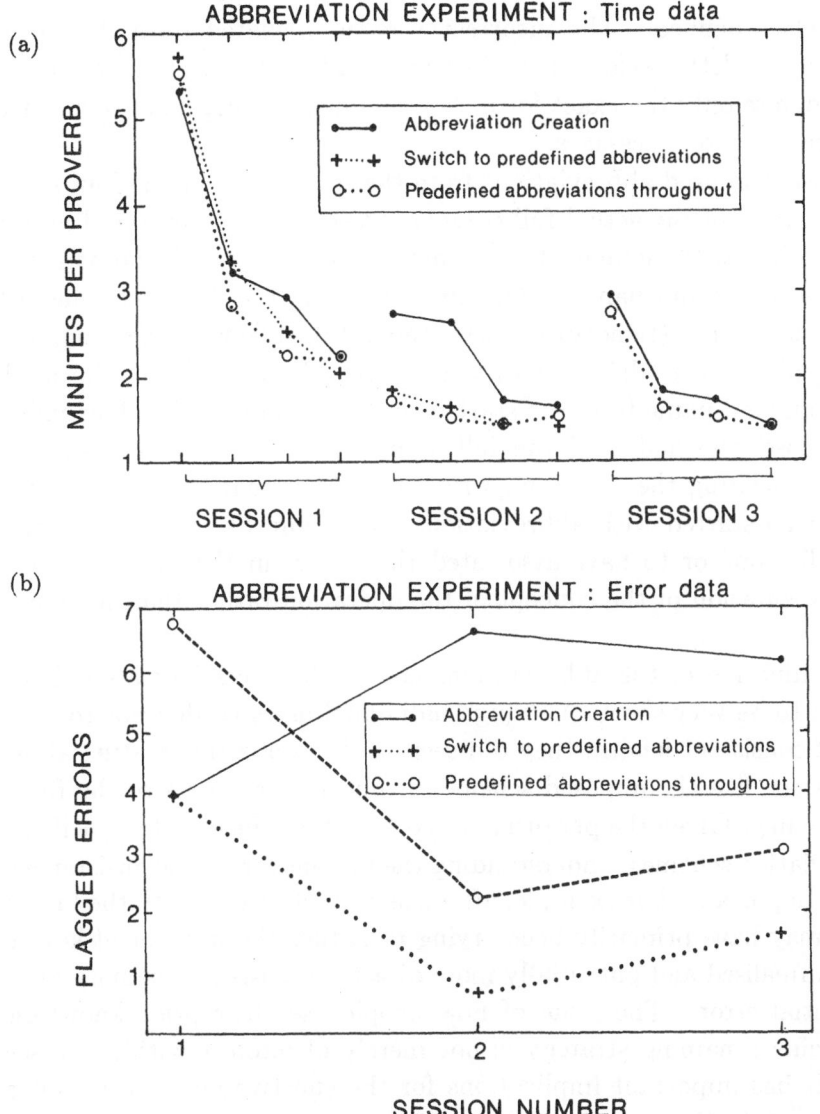

Figure 4.9: Data from Grudin and Barnard (1985): (a) Time data show trend of slower performance in session 2 for abbreviation creation relative to use of the predefined abbreviations and (b) Error data show disadvantage for the predefined abbreviations relative to the full command names in session 1, and a disadvantage for abbreviation creation in the latter sessions. Copyright 1985, by the Association of Computing Machinery Inc, and reproduced by permission

they could not remember them. They tended to generate many plausible abbreviations for the full command name – but not the ones they had themselves defined. The incorrect versions deviated from their original creations in letter selection or in length. Since the system was designed to recognise only the exact form they had defined originally, such variants gave rise to error messages.

Those who used abbreviations from the start rarely entered an incorrect abbreviation of the actual full command names. This group did appear to be entering, with significant frequency, abbreviations of synonyms for the assigned command name. Thus, in attempting the delete command they might enter "re" (remove) or "se" (separate). Likewise they might enter "sp" (split) when "di" (divide) was correct. None of those who used the full command name from the start made this kind of error. This indicated that those who had used the full names for a time had indeed profited from associating the operation with the full command word. In contrast, those who started with abbreviations seem more likely to have forgotten the full name or to have associated the letters in the abbreviation with a representation of the concept of the operation rather than its expanded name.

Examination of the abbreviations created by users indicated that they tended to be very short. Although individual users tended not to generate sets of consistent length, they did seem to be primarily creating abbreviations that, like the imposed scheme, involved short forms of the full command name. Given the performance costs of these inconsistent, self created abbreviation schemes, understanding the reasons for the inconsistencies becomes important. For example, after a session of editing with the full names users may have primarily been trying to reduce the number of keystrokes with unrealised and potentially more disastrous costs for memory and consequential error. The issue of how people use their prior knowledge to determine a naming strategy is not merely of interest within the science base, it has important implications for the effective use of user-tailorable systems. A further two simple experiments were carried out to explore how people create abbreviations, but under circumstances where the amount of relevant background information concerning text-editing was varied.

Three conditions were examined, each of which involved information about command names and text-editing. By contrast with the user group in the previous experiment, who actually did some editing prior to creating their abbreviations, the three conditions involved no active editing. In a paper and pencil task, subjects were asked to generate abbreviations for the specific command names used in the experiment. One group was given a list of command names. The subjects were told that these were commands for computer text editing and asked to generate abbreviations for them. A second group was told in addition that users of text editors found it time consuming to enter full command names and thus that the abbreviations were specifically intended to reduce the number of keystrokes required to enter the commands.

A third group was given more extensive experience of the editing task. These subjects were shown a computer aided demonstration of the ten command operations. The demonstration was self-paced but otherwise passive. At the end of the demonstration, the command operations were presented one by one and the subjects were asked to define abbreviations for them on-line. They too were instructed that typing time was an important factor in command entry. After abbreviating each command, their full sets were presented to them and they were given the opportunity to make changes.

	Abbreviation (Length)	Truncation (Percent)	Vowels (Percent)
Written Instructions	2.47	55	20
Passive Experience	2.36	87	26
Active Experience	1.57	84	27

Figure 4.10: Mean length of abbreviations in characters, the proportion of abbreviations following a truncation rule and the proportion of abbreviation characters that are vowels. The data for the no prior experience and the vicarious experience conditions are combined into an overall mean for written instruction conditions (from Grudin & Barnard, 1984b)

The abbreviations created following written instruction and after passive computer experience could then be compared with those created by the subjects in the original experiment who had active experience of editing proverbs. The characteristics of the sets created under these conditions are shown in Figure 4.10. The two groups whose abbreviations were created on paper did not differ reliably. Hence, they are combined in the figure as "written instructions".

Those who had active experience of text editing create abbreviations that are significantly shorter in length than all the other conditions, even those in which it had been emphasised that typing time was a major factor in command entry. Passive experience of the editing task and the command operations did not lead to any length reduction in relation to the written instruction conditions. The groups that had had either active or passive experience of the editing task produced abbreviation sets with a higher proportion of truncations than was produced after written instruction. Subjects with written instruction were rather more likely to use vowel deletion in their abbreviation strategies. Over the four groups, only three of the twenty-nine subjects produced sets of abbreviations in which all of their abbreviations contained the same number of letters.

Since effects of this sort could easily be related to the particular names used, Grudin & Barnard (1984b) performed a partial replication to check that this candidate empirical regularity would occur with other materials. A set of command names synonymous with the first was constructed and given to two new groups with the same written instructions as before. In addition, another group was given active experience, but less extensive than in the previous experiment. The same overall pattern of results was obtained. Those subjects who created abbreviations after active experience with the text editing task created shorter abbreviations and a higher proportion of truncations than the groups given written instructions and no experience of editing. Once again, less than ten percent of subjects created abbreviations that were of a consistent length.

In creating abbreviations in the absence of editing experience, people generate longer abbreviations and tend to rely more on vowel deletion. This suggests that they viewed the task as one in which they should seek to generate abbreviations that were intended to help reconstruct the full name even when instructions emphasise a need to reduce keystrokes. In contrast, those with some experience of the commands may tend to place less importance on reconstructing the full name but rather more emphasis on producing truncations that discriminate among names. However, only those who have had active experience really seek to minimise the number of keystrokes.

In terms of the candidate empirical regularities for the science base these studies of abbreviation learning and creation of abbreviations confirm and extend findings in the literature. First, as with the Streeter et al study, this series of experiments indicates consistent abbreviation schemes improve learning and performance. Here the inconsistent schemes were user created. A consistent abbreviation rule is, of course, one type of structural description. Hence, the finding is compatible with the same kind of scientific principle called upon to explain the relative advantage of pseudowords over consonant strings in section 3.1 (Grudin & Barnard, 1984a). Ease of learning a particular form of mental representation appears directly related to the availability of a structural description that could facilitate the recovery of information from memory in the early phases of acquisition.

As with the overall extent of ambiguity in name-operation mapping, different abbreviation schemes may also vary in their utility. In the earlier Grudin and Barnard (1984a) study a three letter contraction rule was employed. That set of abbreviations seemed to give rise to rather more extensive performance costs (time and errors) than the two letter truncation rule (errors). The most substantial problems occurred with the inconsistent abbreviation schemes created by users. This evidence again suggests gradations that require a more detailed account. In the earliest stages of learning there is an unautomated mapping within a mental process. In section 3.1, one of the candidate principles proposed that the cognitive system configures to resolve the uncertainty either by inference, direct memory access or by referring to external sources of information. A second principle stated that requirements for inferential processing and/or memory access increase with the overall extent of item uncertainty. Through its effects on memory access the structural description provided by an abbreviation rule is one means by which uncertainty can be reduced. In this respect the two letter truncation rule may have been more readily encodable than the rather more complex three letter contraction rule and hence functioned more readily to reduce uncertainty. Obviously, the users who created their own abbreviations did not produce items that conformed to a consistent structural rule, particularly with regard to length. There would therefore have been far greater levels of uncertainty.

The data also suggested that initial experience of the full command names prior to using abbreviations was advantageous. Use of the truncation scheme throughout also led to different forms of error. Initial experience of using the full command names would, in Barnard's (1987) framework, lead to memory records of those transactions which could subsequently be accessed as one means to resolve uncertainty. This source of knowledge of the full command names would not only assist learning via the presence of a structural description of lexical form (cf. section 3.1), it would also function in the reduction of uncertainty via memory access, to block off those attempts to enter an abbreviation of a synonym of the actual command word (e.g. 're' for remove instead or 'de' for delete).

In contexts where users created their own abbreviations, experiential and strategic factors again came into play. In abbreviation creation, the task may be contextualised as one trading off different components – reconstructing names, discriminating among them and minimising keystrokes. Rarely do users appear to be sensitive to an important feature of abbreviation memorability – that of generating names of consistent length and content.

Interestingly, there are other circumstances in which users make inefficient choices. MacLean, Barnard & Wilson (1985), for example, showed that when users are practiced on alternative methods for data entry into a spreadsheet, there are circumstances where they choose the temporally less efficient method. Young and MacLean (1988) then went on to argue that users simplify the decision space to focus upon only some of the relevant considerations – thereby failing to take into account an element of a method that takes a considerable amount of time. In the present case, users with interactive experience appear to concentrate on the keystroking requirements at the expense of other relevant considerations. Although it will not be pursued here, these kinds of phenomena may be open to interpretation on the basis of a candidate principle for the operation of inferential procedural knowledge. Such a principle might involve a limitation on the number of features that can be taken into account, where the particular features are selected on the basis of their inferred salience to the task, and where precise experience plays a major role in determining salience.

4.3.4   Summary

The purpose of this chapter was to illustrate how a particular experimental paradigm can be utilised iteratively to build up a knowledge base of empirical regularities for interpretation within the science base. The inferred regularities can now be summarised.

The first set of issues explored types of name set that imposed different demands to learn novel meaning and novel form. The data indicated that both kinds of demand led to significant increases in editing time, dialogue transactions and use of the help facilities. The relationship appeared to be systematic. There were performance costs associated with the acquisition of either new form (abbreviations) or new meaning (unrelated command names). There were greater costs associated with learning both new meaning and new form. In spite of these extra performance costs, there was evidence that users could put their knowledge of lexical form to good use. As learning progressed, the pseudoword command names appeared to become lexicalised. Continued high error rates and poor recall suggested that users of consonant strings had considerable difficulty both with the learning of form and with associating these names with the operations to which they referred.

The remaining studies focussed either on the contextualisation of novel meanings for known words (specific and general verbs) or on the creation and contextualisation of novel form (abbreviation schemes). In both cases, the experiments focussed upon demands that more adequately reflected the contrasts that might be found in learning real name sets. The more extreme cases involving entire sets of pseudowords and consonant strings were not pursued further. With the less extreme contrasts, overall measures, such as total editing time, proved less discriminating than more detailed analyses of the ways in which users coped with the precise learning demands.

Where known words needed to be contextualised, it was found that there were only minor differences in the overall time taken to edit proverbs as a function of the semantic specificity of the command name set. Command name sets consisting wholly of semantically specific names were again readily learned. Command name sets consisting of entirely general terms led users to consult a help panel linking command names to definitions of their operations with considerably greater frequency and led to less accurate recall of the operations that the system carried out. Users of the general command name set appeared to be operating a relatively passive strategy in which they consulted help rapidly to resolve ambiguity in name-operation mapping. In contrast, users of the specific name set spent more time actively considering what to do prior to entering a command – a pattern which could well have assisted in learning the systems' operations. These kinds of effects were observed over two conditions of learning – where users had to plan their own editing sequence and when the editing sequence was constrained.

Where name sets consisted of half general and half specific command names, overall performance was very similar to the levels obtained with a set constituted entirely of specific names. There were some minor differences in the ways users coped with the general and specific names within a set. However, the kinds of patterns obtained across experiments involving the contextualisation of known words are perhaps best understood in terms of the overall ambiguity of assigning names to operations. Thus, with specific name sets potential ambiguity in name-operation assignment was low. With the unrelated command name sets used in the first study described, the ambiguity was maximal. Each name could potentially have been assigned to any operation. With the general command name set, potential ambiguity was substantial. Many of the names used could potentially have been reassigned – either within or between pairs.

With the command name sets involving mixtures of general and specific names, the levels of ambiguity were much reduced. Under these circumstances, users could apparently work out what to do almost as readily as they could with the name set made up entirely of specific names. When pairs of related operations were referred to by one specific and one general name, knowledge of the specific member of the pair was apparently sufficient to help users to infer, or otherwise retrieve, the action performed by the general name. When ambiguity was restricted by having two pairs of operations with general names and two pairs with specific names, users were apparently able to cope with the name-operation mappings without great difficulty. The precise strategies adopted for resolving the name-operation mappings were also relative in that usage of the help panels was related to individual differences within the user population.

In the case of learning new abbreviated forms, relative to specific command names, there were performance costs associated with both the short form truncation rule used in Grudin & Barnard (1985) and with the three letter contraction rule used in the earlier study (Grudin & Barnard 1984a). Both rules were structurally consistent. User created abbreviations tended to be structurally inconsistent with even greater performance costs (see also Streeter et al, 1983).

Where abbreviated forms of known words need to be created or learned in novel contexts, users are clearly influenced by their experience of the editing task. Users with no direct experience of the editing task appeared to contextualise the task as one directed towards the reconstruction of the full command names. Those with active experience of editing contextualised the abbreviations for discriminative purposes. Either way, they did not create abbreviations that would have optimised accurate recall. In addition, those users who had some experience of the full command names prior to using abbreviations committed rather different errors than those without such experience. Thus, what is learned is not solely dependant upon the nature of the abbreviation scheme involved.

Naturally, the empirical regularities summarised here may or may not generalise or extend beyond this particular experimental paradigm. However, important criteria for serious consideration within the science base are replicability and that related effects should be systematic. Within each series several of the key effects are replicated. Grudin and Barnard (1984a), for example, performed a replication of the comparison between specific command names and unrelated names. Likewise, the specific and general command name effects occurred over two kinds of task conditions; the absence of a difference between the specific command name sets and the 'mixed' command name sets held for two ways of mixing general and specific commands; and the characteristics of user nominated abbreviations applied over two specific name sets.

The effects also appeared relatively systematic. So for example, increasing the overall extent of ambiguity in name-operation mapping provided relatively consistent effects within and between experiments. Similarly, differences in the abbreviation rule (truncation versus contraction) could reasonably be related to the demands the rules imposed. In contrast to the specific command name set, in both examples, increasing demands led initially to primarily strategic effects on information use, and then to very real costs in time and errors.

4.4 Interpretation within the science base and application of empirical regularities

Given that these text-editing experiments make numerous assumptions in the way in which it simulates the real world of document processing, there are obvious dangers in seeking to apply the results directly to the development of a novel software package. The design requirements of the package are likely to involve very different behavioural constraints from those studied here. In addition, the behavioural regularities obtained with this particular text-editing task form only a subset of those that are known. So, for example, other patterns of data for general and specific command name sets are obtained for task contexts involving the concurrent learning of command names and command syntax (e.g. see Hammond et al, 1987). Likewise, the kinds of command confusion errors obtained are also affected by the conceptual organisation of extended sequences of commands (Barnard et al, 1984). The wider purpose of establishing empirical regularities through experimentation was not simply to provide empirical "facts" but to provide the basis for an appropriate theoretical synthesis within the science base and an effective applications representation.

In describing and discussing the experiments, the kinds of principles that can be derived from this type of research were outlined but not specified in any detail. In particular, the principles made implied reference to a model of the human information processing architecture. Here, we have deliberately avoided the use of technical definitions referring to that particular architecture. Examples of the principles, in their more technical form, are discussed in considerable depth elsewhere (Barnard, 1987). In the introductory sections of this chapter, it was pointed out that one of the main functions of principles in the science base was to support wider generalisation and prediction.

As a general move in this direction principles advocated in sections 3.1 and 3.2 were found to have wider utility in section 3.3. So, for example, the discussion of the apparent lexicalisation of pseudowords, or the advantages of consistent abbreviation schemes, made reference to candidate principles concerning the use of structural descriptions to access memory representations. Likewise, the demands of learning novel form and novel meaning referred to candidate principles concerning requirements to acquire new mappings both between and within separable mental processes.

In order to accommodate these, and other series of experiments, many such principles will be required to provide a fully elaborated science base for naming in the computing context. Indeed, it could be argued that the kinds of principles discussed here could well have been justified from the existing theoretical literature on word forms and 'confusable' semantics. However, as was made clear in the initial treatment of specific and general words, other paradigms could well be called upon to justify alternative principles. A crucial component in operating in an applied paradigm such as computer text-editing is to understand which principles are relevant when cognitive processes are co-ordinated to control extended sequences of action in complex task settings.

To be of real value, any theoretical principles must not only be valid in relation to a broader scientific understanding, they must also be applicable. In this respect, general principles concerning the operation of the human information processing mechanism are notoriously difficult to instantiate in novel circumstances. Given a particular application, many principles may be involved. Determining which principles to use and exactly how to interpret them has often been very much a matter of skilled judgement and experience. In common with others (e.g. Newell & Card, 1985; Polson, 1987), we have assumed that the kind of knowledge in the science representation needs to be recruited to build approximate engineering models which in turn informs interactive system development.

In our case, we have attempted to define a class of approximate model which describes how the human information processing mechanism will cope with the demands of executing dialogue sequences or learning them. These approximate models are referred to as cognitive task models. In Long's terminology, they are our "applications representation." Their construction requires making the underlying principles explicit together with a set of assumptions that permit the principles to be applied in the context of human-computer dialogues. The assumptions required here can differ substantially from those that would be required to build a formal theory within the science representation itself. In addition, the very complexity of the process of applying scientific understanding to real world tasks argues strongly for the use of artificial intelligence techniques that can potentially manage the problem of intersecting the many different principles that could be relevant to the prediction of user performance.

The actual building of cognitive task models has recently been explored by formalising rules for modelling user cognition in prototype knowledge bases for an expert system (see Barnard et al, 1986; 1987). The knowledge bases incorporate rules for building a four component description of the operation of the human cognitive mechanism. The components define configurations of mental processes required by particular task demands; the status of knowledge proceduralisation for each mental process within a configuration (such as the number of processes for which proceduralisation exists); the properties of memory representations likely to be accessed during learning and performance (such as the degree of item uncertainty in those representations); and a description of the way in which the entire cognitive system is co-ordinated and controlled. These descriptions are derived from the kinds of principles outlined earlier and separate knowledge bases are used to infer the probable aspects of user behaviour.

So far, knowledge bases have been constructed that effectively replicate the reasoning underlying the explanation of a limited range of experimental phenomena. These knowledge bases are, in their current form, capable of generalising to make predictions to settings other than those which gave rise to the principles. For example, the knowledge base representing principles derived from studies of command names, can, with a small number of additional assumptions be generalised to iconic forms of dialogue (e.g. see Barnard et al, 1987).

The predictive validity of this approach to building an applications representation on the basis of purely experimental research has yet to be established. The adequacy of the modelling relies on the assumptions used in designing experimental simulations, on the replicability of assumed empirical regularities and upon theoretical synthesis capturing the appropriate principles. It is our belief that the chances of success will be considerably enhanced if such models are firmly grounded on a well differentiated understanding of user performance: an understanding established through the cumulative acquisition of evidence from different experimental paradigms.

References

Baddeley, A. D. (1966). Short term memory for word sequences as a function of acoustic, semantic and formal similarity. *Quarterly Journal of Experimental Psychology*, **18**, 362-365.

Barnard, P. (1985). Interactive cognitive subsystems: A psycholinguistic approach to short term memory. In *Progress in the Psychology of Language*, vol 2, chap 6, ed. A. Ellis, pp.197-258, London: Lawrence Erlbaum Associates.

Barnard, P. (1987). Cognitive resources and the learning of human computer dialogs. In *Interfacing Thought: Cognitive Aspects of Human-Computer Interaction*, Chap 6, ed. J.M. Carroll, pp.112-158, Cambridge Mass: MIT Press.

Barnard, P. (1988, in press). The contributions of applied cognitive psychology to the study of human-computer interaction. In *Human Factors for Informatics Usability*. ed. B. Shackel & S. Richardson, Cambridge: Cambridge University Press.

Barnard, P. & Grudin J. (1988). Command names. To appear in: Helander M. (ed) *Handbook of Human-Computer Interaction*. Amsterdam: Elsevier Science Publishers BV (North Holland), pp.237-255.

Barnard & Hammond, (1982). Usability and its multiple determination or the occasional user of interactive systems. In *Pathways to the Information Society*, ed. M.B. Williams, pp.543-548, North Holland: Oxford.

Barnard, P., Hammond, N., MacLean, A. & Morton, J. (1982). Learning and remembering interactive commands in a text-editing task. *Behaviour and Information Technology*, **1**, 347-358.

Barnard, P., MacLean, A. & Hammond, N. (1984). User representations of ordered sequences of command operations. In *Interact 84*, vol 1, ed. B. Shackel, pp.434-438.

Barnard, P., MacLean, A. & Wilson, M. (1988). Navigating integrated facilities: terminating and initiating interaction sequences. In *Proceedings of CHI '88* (Washington, May 15-17th), New York: ACM.

Barnard, P., Wilson, M. & MacLean, A. (1986). Approximate modelling of cognitive activity: A concept demonstrator for an interactive design tool. *IBM Hursley Human Factors Report*, HF 123, Sept 1986, p.37.

Barnard, P., Wilson, M. & MacLean, A. (1987). Approximate modelling of cognitive activity: towards an expert system design aid. In *CHI + GI '87 Human Factors in Computing systems and Graphics Interface*, (Toronto 5th-9th April 1987), pp.21-26, ACM: New York.

Black, J., & Moran, T. (1982). Learning and remembering command names. In *Proceedings of Human Factors in Computer Systems* (Gaithersburg), pp.8-11, New York: ACM.

Broadbent, D. E., Cooper, P. F., Fitzgerald, P., & Parks, K. R. (1982). The cognitive failures questionnaire (CFQ) and its correlates. *British Journal of Clinical Psychology*, **21**, 1.

Carroll, J. M. (1978). Names and naming: an interdisciplinary Review. *IBM Research Report* RC 7370.

Carroll, J. M. (1982). Learning, using and designing command paradigms. *Human Learning*, **1**, 31-62.

Craik , F. M. & Lockhart, R. (1972). Levels of processing: A framework for memory research. *Journal of Verbal Learning and Verbal Behavior*, **11**, 671-684.

Ehrenreich, S. L., & Porcu, T. A. (1982). Abbreviations for automated systems: Teaching operators the rules. In *Directions in Human-Computer Interaction*, ed. A. Badre, & B. Shneiderman, pp.111-135, Norwood, NJ: Ablex.

Grudin, J. & Barnard, P. (1984a). The cognitive demands of learning and representing command names for text-editing. *Human Factors*, **26** (4), 407-422.

Grudin, J. & Barnard, P. (1984b). The role of prior task experience in command name abbreviation. In *Interact '84*, ed. B. Shackel, pp.439-443, Amsterdam: Elsevier.

Grudin, J. & Barnard, P. (1985). When does an abbreviation become a word? And related questions. In *Proceedings of CHI '85 Human Factors in Computing Systems* (San Francisco), pp.121-125, New York: ACM.

Gumenik, W. E. (1979). The advantages of specific terms over general terms as cues for sentence recall: instantiation or retrieval. *Memory & Cognition*, **7**, 240.

Hammond, N., Barnard, P., Clark, I., Morton, J., & Long, J. (1980). Structure and content in interactive dialogue. *IBM Hursley Human Factors Report*, HF034.

Hammond, N., Barnard, P., Morton, J., Long, J., & Clark, I. (1987). Characterising user performance in command-drive dialogue. *Behaviour and Information Technology*, **6**, 159-205.

Hirsh-Pasek, K., Nudelman, S. & Schneider, M. L. (1982). An experimental evaluation of abbreviation schemes in limited lexicons. *Behaviour and Information Technology*, **1**, 359-370.

Hodge, M. & Pennington, M. (1973). Some studies of word abbreviation behaviour. *Journal of Experimental Psychology*, **98**, 350-361.

Landauer, T. K. & Galotti, K. M. (1984). What makes a difference when? Comments on Grudin and Barnard. *Human Factors*, **26** (4), 423-429.

Landauer, T. K., Galotti, K. M. & Hartwell, S. (1983). Natural command names and initial learning: A study of text editing terms. *Communications of the ACM*, **26**, 495-503.

Ledgard, H., Singer, A., & Whiteside, J. (1981). *Directions in human factors for interactive systems.* Lecture Notes in computer science 103, Berlin: Springer Verlag.

Long, J., Hammond, N., Barnard, P., Morton, J. & Clark, I. (1983). Introducing the interactive computer at work: the users' views. *Behaviour and Information Technology*, **2**, 39-106.

Mack, R. L., Lewis, C., & Carroll, J. M. (1983). Learning to use word processors: problems and prospects. *ACM Transactions on Office Information Systems*, **1**, 254-271.

MacLean, A., Barnard, P. & Hammond, N. (1988). Nameset constitution and text-editing performance: mixing general and specific command names within a set. *IBM Hursley Human Factors Report*, HF 138.

MacLean, A., Barnard, P. & Wilson, M. (1985). Evaluating the human interface of a data entry system: user choice and performance measures yield different trade-off functions. In *People and Computers: Designing the Interface*, ed. P. Johnson & S. Cook, pp.172-185, Cambridge: Cambridge University Press.

Newell, A. & Card, S. K. (1985). The prospects for psychological science in human-computer interaction. *Human-Computer Interaction*, **1**, 209-242.

Norman, D. A. (1981). The trouble with Unix. *Datamation*, **27** (Nov), 139- 150.

Polson, P. G. (1987). A quantitative theory of human-computer interaction. In *Interfacing Thought: Cognitive aspects of human- computer Interaction*, ed. J. M. Carroll, Chap 8, pp. 184-235, Cambridge Mass: MIT Press.

Rosenberg, J. (1982). Evaluating the suggestiveness of command names. *Behaviour and Information Technology*, **1**, 371-400.

Rosenberg, J. & Moran, T. (1985). Generic Commands. In *Human-Computer Interaction – Interact '84*, ed. B. Shackel, pp.245-249, Amsterdam: North Holland, Elsevier.

Scapin, D. L. (1981). Computer commands in restricted natural language: Some aspects of memory and experience. *Human Factors*, **23**, 365-375.

Streeter, L. A., Acroff, J. M. & Taylor, G. A. (1983). On abbreviating command names. *The Bell System Technical Journal*, **62**, 1807-1826.

Thios, S. J. (1975). Memory for general and specific sentences. *Memory & Cognition*, **3**, 75.

Young, R.M., & MacLean, A. (1988). Choosing between methods: Analysing the users decision space in terms of schemas and linear models. In *Proc. CHI '88 Human Factors in Computing Systems* (Washington, May 15-19th), New York: ACM.

5

A notation for reasoning about learning

Stephen J. Payne

5.1   Introduction

One of the most ambitious aims of cognitive science is to specify the general *functional architecture* of human cognition: the processing resources of the mind which are common to all individuals and which enable and constrain human performance on all cognitive tasks. An early attempt to express such an architecture in the form of a programming language was Newell's production system (1973a). Notable recent proposals have been made by Anderson (1983), Laird, Newell & Rosenbloom (1987) and, in a rather different style, by the "new connectionists" with parallel distributed processing models (eg Rumelhart & McClelland, 1986).

Such an ambitious aim is unlikely to be achieved in the forseeable future. Yet cognitive ergonomics cannot afford to wait for the dust to settle in the scientific arena and reveal a complete, exact theory of the human mind. How might the immediate engineering goals of cognitive ergonomics be informed by the ongoing scientific enterprise of modelling cognitive architecture, with all its strengths and weaknesses?

The greatest strength of architecture-level theories is that they provide a common language for the expression of algorithm-level theories of separate tasks. This is as much a merit for the applied problems of cognitive ergonomics as it is for the problems of pure science. In cognitive psychology it serves to avoid the ad hoc aspects of the one-task-one-theory approach, and to manage the explosion of empirical phenomena (Newell, 1973b). In cognitive ergonomics, where the task is the use of devices to achieve some goal, a common language for theorising about device use might enable us to compare directly the psychological properties of different devices, perhaps in advance of their being built, and certainly without expensive empirical trials.

The use of cognitive architectures as common languages in this way goes one better than attempting to adapt the kind of formal descriptions typically used throughout computer science (eg in programming language definition) to problems in cognitive ergonomics. Although such non-psychological formalisms allow some rigorous reasoning about design changes, they do not allow comparison of the psychological implications of design changes, except indirectly. For example, Reisner (1981), in her pioneering work, was only partly successful in reasoning about learnability from BNF descriptions of a user interface (see Green, 1980).

A second strength, more heuristic in nature, can derive from adopting explicit, universal architectures for modelling cognition. By forcing the analyst to express complex task performance in some pre-determined, restrictive language an architectural theory can help the analyst to dig out the structure of the task. Just as the programming of computers can act as a "discipline to the imagination" of the scientific theorist, or even as a complex rhyme scheme can prompt the poet to new expressive insights, the constraint of a formal language might yield deeper understandings to the task-analyst.

The weakness of general purpose cognitive architectures is, paradoxically, that they need to be computationally very powerful. The main problem with computational power derives from the status of architectures as theories. Architectures do not make empirical predictions directly, but through the expression of some particular algorithm for performing a task. The architecture only contributes empirical muscle to the extent that it favours the particular algorithm over potential competitors. But the more powerful the architecture, the more able it is to express alternative algorithms, so that more power generally means less empirical force; architectural theorists are often faced with a trade-off between explanatory and empirical adequacy.

This need not mean that powerful architectures are empirically barren, for although they can make few predictions about what can or cannot be computed, psychology will often happily rely on chronometric predictions, and even complete Turing machines cannot necessarily support algorithms of particular time-complexities. In practice, however, cognitive architectures gain validation only indirectly and imperfectly, in that they provide convenient ways of expressing certain algorithms, making these algorithms seem more "natural" than their competitors. Such arguments are easier to construct post hoc, starting with the behaviour that a candidate algorithm should generate. This has meant that architectures are more typically supported by showing how well they capture existing empirical data, rather than by testing any novel empirical predictions that they generate. This state-of-the-art is well exemplified in the research on parallel distributed processing, where the current focus appears to be on constructing models which capture well-known empirical phenomena.

I do not wish to raise any deep philosophical objection to this approach (although others might!), merely to question its utility in cognitive ergonomics, where interest is naturally more focussed on prediction than on explanation, so that capturing data for its own sake looms small, and coarse predictions about large differences in learnability and usability of different designs are more important than are finer predictions of response times and the like.

Further, the lack of constraint means that for their second potential benefit, as insight-prompting disciplines, existing cognitive architectures may be *less* well suited than the formal notations of the computer scientist, which, like those of the computational linguist are often very deliberately limited to the weakest computational mechanism possible.

Weighing the strengths and the weaknesses suggests a particular approach to cognitive ergonomics. By focussing on a narrow domain of cognitive activity it may be possible to propose cognitive architectures that sacrifice scope of application so as to constrain computational power and thus gain empirical force and insight-prompting discipline.

The major difficulty facing this strategy is the richness of the task of computer-use – it is hard to imagine an aspect of cognitive performance which is *not* implicated in the use of computer systems. To avoid this difficulty it is necessary to divide the task into manageable portions.

The research project on which this chapter provides a window focusses on the analysis of learning and learnability. Why is it that some systems are easier to learn than others? Why do some users adopt inefficient methods for performing tasks? How might systems and instructions be designed to foster efficient learning? To answer questions like these we must analyse what users *need to know* to use the system successfully: we can concentrate on addressing competence and its development, without worrying too much about the details of performance.

Following the strategy outlined above, a cognitive architecture is advanced which allows us to express the knowledge needed to use computer systems in a way that reflects important properties of the mental representation of such knowledge. The architecture models users' knowledge of task languages – the communication channel through which users instruct the machine to perform tasks. Task languages present one of the learner's most severe cognitive challenges. Systems which use hard-to-remember command sequences, strange names, or meaningless abbreviations are notoriously unpopular, and may some day be extinct, but even WIMP interfacing (which uses Windows, Icons, a Mouse, and Pull-down menus) can involve complex structures, as we shall see.

A "cognitive architecture" which is constrained to modelling competence in task languages seems hardly worthy of such a grand name. For such a pragmatically limited architecture, I prefer the term "notation". The notation may be applied to any particular task language to provide a model of competence in that task language.

In the terminology of Long's framework of Chapter 1, a psychological notation allows us to map the science base onto an application representation within a single theory. A psychological notation expresses limitations and capabilities of the cognitive system in some task-domain, thus expressing psychological theory from the science base; the use of a psychological notation to model competence in a particular task is intended to play the role of an application representation, allowing designers insight into the learnability of system designs.

The notation advanced in this chapter is TAG, a notation for expressing task-action grammars, which model users' competence. TAG has been described elsewhere (Payne, 1985; Payne and Green, 1986), so this chapter has more particular goals than a fresh espousal of the model. These goals are:

- To emphasise TAG's potential as a notation for reasoning about learning. TAG's ability to make coarse empirical predictions about the relative learnability of different languages does not figure in this chapter (but see Payne and Green, 1986; Payne and Green, in press). Instead I focus on the use of TAG as a tool for analysing the variety of possibly perceived structures in a task language, and the implications of these structures for learning.

- To advance TAG's capability by considering a model of the way users construct device-oriented problem spaces, and the implications of this model for task-action grammars. In particular, I will argue that users with richer conceptual models of the device can build more efficient task-action grammars.

These goals will be addressed through a TAG analysis of a mouse-driven text-editor, the Xerox TEDIT system. This case study might, I hope, help achieve a third goal, to instruct would-be analysts how to build task-action grammars of real systems, ie to provide them with a method for building models using the TAG notation. The arguments made in the chapter will rely on intuitive force, rather than empirical evidence. In several places the case analysis does lead to firm empirical predictions, but these have not yet been tested.

Before we can tackle TEDIT, a brief introduction to the fundamentals of task-action grammars is necessary.

5.2   Task-action grammars

5.2.1   The TAG notation

If there is one feature of task languages which common wisdom suggests affects learnability it is *consistency* (see Payne and Green, 1986 for a discussion). Experimental results in the literature of human-computer interaction bear this out, and exhibit the learnability advantages of several organisational principles. For example: the choice of lexical names which reflect semantic relations between their referent operations (Carroll, 1982; Green and Payne, 1984), or which possess redundant terms to reflect the structure of the command language (Scapin, 1982); the design of command language syntax that places commonly occurring elements in a consistent position, or that mirrors the semantic organisation of the task world (Barnard *et al*, 1981; Payne and Green, in press).

It turns out that the advantages of all these separate aspects of consistency can be explained by two assumptions about the way users represent task languages: they categorise the world of tasks according to the similarities and differences among tasks, and they perceive higher-order rules which utilise the semantics of the task-world to capture structural resemblances between separate task-action mappings.

The main predictive power of TAG as a psychological notation results from the fact that it embodies these assumptions. The compactness and simplicity of the TAG notation results from the fact that the assumptions can be embodied by a generative grammar with a few special notational devices.

The notation will be introduced through just one simple example, illustrating only the major notational devices. Some of TAG's other notational devices will be used in the analysis of TEDIT; these will be explained one by one as they arise.

One of the experimental text-editing languages used by Green and Payne (1984) contained four commands for moving the cursor. To move the cursor forward by a word or a character, the "control" key was used to modify a single letter code denoting the unit of travel – "W" for word, "C" for character. To move backwards by the same amounts the "escape" key was used as a modifier.

Let us describe this tiny task language fragment to consider how TAG supports the psychological assumptions we noted above.

i Users represent the world of tasks according to the similarities and differences between tasks.

TAG identifies "simple-tasks" that the user can perform routinely, and that require no control structure. These tasks appear to have a special status as far as learnability goes, because tasks involving "programs" of more primitive tasks can be performed spontaneously by users who have only learned the primitives (Douglas, 1983). A TAG description of a task language lists a dictionary of all the simple-tasks, defining each by a collection of semantic components (valued features), in particular using features which distinguish between the meanings of the simple-tasks. (Sometimes it aids exposition to display redundantly the features that are used and the possible values each can take. This is done in the more extended case analysis below.)

For the current language fragment, the featural definition of the four simple-tasks seems straightforward (see Payne, 1985).

> move cursor forward a character {Direction=forward,
> Unit=char}
> move cursor back a character {Direction=back, Unit=char}
> move cursor forward a word {Direction=forward,
> Unit=word}
> move cursor back a word {Direction=back, Unit=word}

The two features, Direction and Unit clearly express the similarities and distinctions among the four commands. If the wider task language were to be considered, additional features would label these four simple-tasks as cursor movement. For example, if the other tasks all either inserted or deleted text, then a three-valued feature Role=insertion, Role=deletion, Role=cursor-movement, would distinguish the separate classes.

Choosing the features with which to represent tasks in the simple-task dictionary might not always be as straightforward as it is in this limited example. TAG does not offer any prescription for this specification, the featural description of tasks is an input to the task-action grammar that must be defined by the analyst. The basic analytic technique is to choose the minimal set of features that captures the salient similarities and distinctions within the group of simple-tasks; the analyst's skill must be used to find features that are psychologically salient to the user.

This approach is reasonable for the coarse analysis of relative learnability, for it turns out to be quite restrictive – as we have just seen, features are only included if they describe distinctions within the group of simple-tasks. But if we extend our ambitions to include a more thorough analysis of the learning problems posed by devices and especially if we wish to use TAG as a basis for a dynamic learning theory which can contribute to the design of instructional systems, then a more refined approach would be preferred. One possibility is to derive the featural definitions empirically, using some psychological scaling technique. This would have the added advantage of being sensitive to possible individual differences. Such extensions are outside the scope of this chapter.

ii Users perceive higher-order rules which use the semantics of the task world to capture structural resemblances between task-action mappings.

TAG models the perception of higher-order regularities by using feature-tagged rule schemas to rewrite tasks into actions. The top-level rule schema for the four-command example is:

EG1 Task [Direction, Unit] → symbol [Direction]
+ letter [Unit]

(The rules are numbered EGi for "example rule i" to aid later references.)

Such top-level rules, which rewrite the "Task" symbol (which is reserved exclusively to label such rules), have a special status in task-action grammars, for they display the overall structure of command sequences, and are therefore taken to be an especially important index of perceived complexity. Payne and Green (1986) argue that when comparing two similar task languages, a count of top-level rules is the best index of learnability.

Despite their special status, the top-level rules work in exactly the same way as all other rules. The grammatical symbols *outside* the square brackets have no semantic status – "symbol" and "letter" are simply convenient labels, and could just as well have been "label1" and "label2". The features do have a semantic status of course, and have already been seen in the dictionary of simple-tasks. As features contain the *only* semantic information in TAG rewrite rules, TAG may be seen to embody a simplified model of task language semantics, in which the semantics of a sentential command is derived additively from the semantics of its components.

Rules are expanded by assigning values to the features inside the square brackets. A feature that appears more than once in the same rule must be assigned the same value throughout for a valid expansion. So one expansion of rule **EG1** would be:

Task [Direction=forward, Unit=char] →
symbol [Direction=forward]
+ letter [Unit=char]

There are of course four separate expansions of this kind, each having the status of an ordinary grammatical rewrite rule: the featural device allows higher-order rule-schemas to capture the regularity between similar lower-order task-action mapping rules. When the features have been assigned values, the Task token denotes a unique simple-task, in the example "move cursor forward by a character". Any other tokens that are not terminal symbols must be rewritten elsewhere in the complete grammar. To complete our example grammar we need a few more rules:

> EG2 symbol [Direction=forward] → "CTRL"
> EG3 symbol [Direction=back] → "ESC"
> EG4 letter [Unit=char] → "C"
> EG5 letter [Unit=word] → "W"

Double quotation marks denote the terminal symbols of the grammar. These are taken to be "action specifications" which, in a full performance theory would need to be passed on to some action interpreter. Although TAG does have some notational devices for modelling the mnemonic appropriateness of the "C" and "W" abbreviations, we will not consider such naming rules in this chapter (see Payne and Green, 1986, secton 2.3).

This limited example expresses the central ideas behind task-action grammars. Before attacking a much larger and more elaborate task language, I wish to point out some limitations of the model, and move towards an extension.

5.2.2   *Extending TAGs with conceptual models*

One weak point of the task-action grammar model concerns the specification of simple-tasks. Simple-tasks play the role of operators in the classical problem space/search model of problem solving (Newell and Simon, 1972) – ie they are the primitive operations from which more complex goal-driven behaviours are built; they also determine the nature of the mapping function from tasks to actions that defines the system's task language for the user.

The first problem is: for a given system, what are the simple-tasks? Payne and Green (1986) offer constraints on simple-tasks – in particular they must not involve control structures – but offer little help to the analyst, other than to cite Moran's (1983) ETIT analysis and argue that simple-tasks are "internal" in Moran's sense: they are determined in part by the concepts internal to the system, rather than in the external task world. So for a cut and paste editor which makes no distinction between words, sentences and arbitrary character strings, an operation like "delete string" is a simple-task.

I suggest an approach to the definition of simple-tasks, derived from a model, currently under development, of the way users construct a device-centred problem space. Although a computational analysis of this issue is under development (Payne, 1987a; Payne 1987b), our current purposes are restricted to illustrating the implications of this approach for the analysis of task language learnability; a brief informal treatment will suffice.

Following Moran (1983), it is hypothesised that users need to model two separate conceptual worlds – the "external" world of goals and subgoals, the "internal" world of the system – and the relationship between them. The psychological model of each of these two worlds may be regarded as a state space. So a device-user's problem space must contain two separate state spaces, and the semantic mapping between them. Payne (1987b) calls this the "yoked state space" hypothesis. Of prime importance in the analysis is the ontology of each state space – the "conceptual entities" (Greeno, 1983) from which it is built.

Although the user's goals are defined in the external world (the "goal space"), he or she must solve problems by applying operators in the internal world (the "device space"). At the very least then, the entities in a user's device space must be capable of representing entities in the goal space. For a simple cut-and-paste text-editor, this minimal device space may include the concept of a text *string*, which maps onto all the text-objects in the goal space. For a text-editor which has special operations for dealing with characters, words, sentences and so on, these text-objects are themselves entities in the minimal device space. "Minimal" device spaces, then, are conceptual models of the device which allow all the objects and relations in the user's goal space to be mapped onto the device.

Construction of a minimal device space is adequate for task performance, but it is often sub-optimal. Consider the commonplace operation of marking a piece of text, and copying it to a new position. The "marking" part of this operation has no meaning in terms of the minimal device space, it merely plays a syntactic role in the structured "copy" operator. This "operational account" of the mark operation restricts the problem solving flexibility of the user, who cannot infer, for example how to make multiple copies of the same marked-item, because he or she *has no concept of a marked item.*

Problem solving will be facilitated if the user instead adopts a "figurative account" of the mark operation, by elaborating the minimal device space to give the operation an independent meaning. For example, the user might construct a "buffer" model, and construe marking a text-string as copying it into the buffer.[1]

As a second example, consider the commands that might exist in a

---

[1]The terminology of "operational" and "figurative", is intended to be suggestive of the major resources underlying the construction of each kind of model. "Operational" accounts are constructed by focussing on methods for operating the device. "Figurative" accounts are constructed in order to understand the device, and will rely heavily on analogies to more concrete, or empirically familiar, notions.

keyboard-driven screen-editor for moving the cursor (such as the commands analysed above). Moving the cursor has no autonomous effect on the goal space so that the cursor need not be part of the minimal device space, and the move-cursor commands could be given operational accounts. Deleting a word, for example is achieved by moving the cursor to that word then typing the delete-word key sequence. For some reason this seems very unlikely – it is hard to imagine a user of a screen editor not treating the cursor as a conceptual entity in the device space with a meaning of its own, so that moving the cursor is to change the state of the device, irrespective of what task is performed subsequently. Why is this intuition so strong? There are at least five crucial attributes of cursors (which distinguish cursors from copy-buffers) that predispose users to construct figurative accounts of cursor operations. The point of describing each of these attributes in turn is that they may give us general clues about the way device attributes influence the construction of conceptual models.

i First, the cursor is visible on the display. Perceptual properties of the device will have a large impact on the user's conceptual model of the device, and one compelling generalisation is that visible objects will become part of the device space whenever the user can construct some figurative meaning for them.

ii In the case of the cursor, this construction is facilitated by its second important property. The concept of a "point of action", which provides a meaning for the cursor is a very available concept, being part of the ontology of type-writers (the head) and of paper-and-pencil text production (the pencil-point).

iii Thirdly, the move-cursor operations, if given an operational account, would appear in almost all simple-tasks. One way in which figurative accounts may be derived is through "deconstructing" or analysing methods into their component operations, and constructing an independent sense for each component (Payne 1987b). Noticing common elements of separate task-action mappings may facilitate this deconstruction.

iv In many circumstances, the move-cursor component of a task will itself require problem solving – finding a sometimes quite elaborate sequence of actions which move the cursor to its goal position. In these cases, including cursor-movements as components of other simple-tasks would mean embedding an entire problem solution within a single problem-solving operator.

v Finally, the cursor is typically referred to by name in both formal training manuals and informal tutorials given by expert friends. Verbal labelling of objects may be a very general influence on the construction of device models.

Unfortunately, all these arguments currently rest on intuitive, rather than empirical evidence. To strengthen this intuitive base, consider a text editor in which the cursor does not possess some of these attributes, and for which an operational account of cursor movements seems less counter-intuitive. Mouse driven text-editors use a pointer, which represents the mouse-position, as well as a cursor. The cursor is moved by moving the pointer and then clicking a mouse button. Of the five prompts to the con-struction of figurative accounts, two that are less prominent in mouse- than in keyboard-driven editors are **iv** and **v**. Cursor-movement with a mouse becomes very straightforward, not requiring problem solving in any mean-ingful sense, merely a point-click action. For this reason, perhaps, substan-tially less attention can be given to the cursor in instructional materials.

In these mouse-driven editors, one can indeed imagine users perceiving a point-click-do structure for most commands, which gives an operational account to cursor movement. Such a construction would become even more likely if the cursor were invisible. None of the operations would be made impossible by such a peculiar design and many would be no more awkward. Yet, as in the buffer example, inefficient strategies might result: if the mouse (and pointer) move accidentally, for example, users may well point again, redundantly, to an unmoved point-of-action.

So, if our analysis of the cursor can be extended, sometimes users will be offered figurative accounts of operators by the instructions they receive, and sometimes the device will be designed in a way that figurative accounts are strongly suggested (as a second example, the MacWrite text editor has a visual "clipboard" to serve as the copy buffer). But often, I conjecture, the construction of figurative accounts will be a learning difficulty for users, so that the transition from operational to figurative accounts may be one important characterisation of expertise (cf DiSessa, 1986).

Whether an operation is given a figurative or an operational account has implications for the TAG description of how that operation is effected in the device's task language. If the operation is given a figurative account it has independent status as a problem-space operator, ie it is a simple-task. If however it is given an operational account then it will appear in a TAG description only as a non-terminal symbol. Furthermore, the features available to categorise the world of simple-tasks change according to the degree of elaboration of the device space, for the goal space and the device space afford the only source of semantic features. In our example above, whether a simple-task operates on the buffer or not is one possible semantic feature that is only available to users who have constructed an elaborate device space. So an elaborate device space may lead to a restructuring of the perceived task language, as well as the grosser change arising from the deconstruction of simple-tasks.

For any given user at any given time, the device space will offer a mix of figurative and operational accounts of operations. But to analyse the implications of device space elaborations for task language learnability it is useful to consider the two extremes – a minimal device space, and a device space which gives figurative accounts to all potentially autonomous operations.

In the next section we provide TAG analyses of TEDIT, a Xerox workstation text editor. The analyses illustrate the interrelationship between the conceptual model and the task language, by providing alternative TAG analyses for both a minimal and a fully elaborated device space. Although a major focus of the work on TAG has been to provide simple metrics for the assessment of relative learnability between different task language designs, in the following example the focus is on the use of TAG as a notation for reasoning about learning: to give designers insight into the structural properties of the task language, and their implications for the learner/user.

5.3   A case analysis

TEDIT is an advanced text-editor developed at Xerox PARC. It uses "WIMP" technology, and runs in the InterLisp environment on Xerox D-machines. TEDIT provides users with a great many facilities, of which I will only consider the "core" text-manipulation operations: insertion, deletion, replacement, movement and copying of text-units. The task language to accomplish these operations it is quite convoluted to explain verbally; nevertheless, I shall try.

To insert new text in TEDIT, the user points to the desired location, clicks either the left or middle mouse-button, and then types. A left-button click selects a character, and the nearest inter-character space – either before or after the character, depending on the exact position of the pointer. Inserted text squeezes into this inter-character space, extending it appropriately and reformatting appropriately. A middle-button click selects an entire word, and the nearest inter-character space immediately before or after the word, into which any inserted text will go. (It is also possible to select other text units, such as a line or a paragraph, or to point to the end of the file, by pointing the mouse to special positions, but including these specialisms in the analysis would make it top-heavy for little gain.)

If, instead of typing from the keyboard the user presses the DEL key after making a selection with the mouse, the character or word that has been selected is deleted. It is also possible to delete an arbitrary string of characters using the DEL key. The string must first be selected by "extending" the initial selection – pointing with the mouse to the right hand extreme character of the required string and clicking the right button. It is important to note, however, that extended selections behave differently from character or word selections. They are removed *whatever* the subsequent key-stroke. They can be described as in the state "pending removal". So new typing *replaces* the extended selection.

To copy text to a new location an operation called "Shift-select" is used. Having already marked the location to which the text is to be copied, "SHIFT" is held down, and the text-unit to be copied is selected: using the left button for a character, the middle button for a word, and extending to an arbitrary string with the right button. This method of copying uses a reversed syntax from many editors, in which the to-be-copied text is specified *before* the location. Moving text is just like copying it, except that "SHIFT" *and* "CONTROL" must be held down while the text is being selected. Initial text selections will also be deleted if "CONTROL" is depressed during selection.

If the location for a move or a copy is marked by an extended selection, then the selected text is replaced by the moved or copied text, on release of the control- and shift- keys. In effect then, there are two special operations: move-replace and copy-replace.

Those then are the tasks and actions which define the TEDIT task language. To analyse the learnability of this system I will develop two separate task-action grammars: one which assumes the user has only constructed a minimal device space (let us call that the Minimalist's TAG) and a second which assumes the user has elaborated the device space to offer figurative accounts for as many operations as possible (Elaborator's-TAG).

5.3.1   *Minimalist's TAG for TEDIT*

The minimal device space for TEDIT clearly needs to include both special text-objects, like character and word, and the general notion of a text string, as each of these is the meaningful unit for an available action. The cursor, which is moved with the mouse and used for a variety of pointing operations will not be considered a part of the minimal device space, for the reasons discussed above. (The nature of a not-quite-minimalist TAG for users who construct the cursor as a conceptual entity will be discussed below.)

Other than these, however, no special conceptual entities are needed. The simple-tasks then, will be any operations which affect these objects, which are the ones that we have distinguished above, namely: insert text before or after a character, or before or after a word; delete character, word, or string; replace string; copy character, word or string to before or after a character or word; move character, word or string to before or after a character or word; copy-replace string with character, word, or string; move-replace string with character, word, or string. So there are thirty eight simple-tasks in all; what are the features which categorise them?

There are several clear similarities and distinctions between tasks which our featural analysis should reflect. It is trivially clear, for example, that delete word, delete character and delete string only differ from each other in terms of the text-unit operated on. Replacing a string should have a combination of the features of insertion and deletion. Both copying and moving are related to insertion, but differ in that the text-unit to be inserted already exists. Copying and moving are themselves very similar, but differ from each other according to whether this "source" stays in place or disappears. And so on.

To capture these interrelationships in featural terms, it is convenient to consider the effects of the operations on a target and a possible source (Moran, personal communication). The following features suffice:

Feature	Possible values
Target.Insert?	yes, no
Target.Delete?	yes, no
Target.Unit	character, word, string
Source.Exist?	yes, no
Source.Remove?	yes, no
Source.Unit	character, word, string
Before/After	before, after, any

Obviously there are some cooccurrence restrictions on these features. If Source.Exist takes the value no, Source.Remove is not a feature; the Before/After feature applies only to text insertion, moving or copying tasks. In an analysis of larger systems it may be worth explicitly noting such dependencies.

My conjecture then, is that novice users with completely unelaborated device spaces may represent the semantics of the task world in the way described in table 5.1 It is worth mentioning that in this table a new typographical convention for the expression of feature-sets has been adopted. In previous task-action grammars, including the TAG for cursor movement in section 5.2, standard set-theory notation of curly-brackets has been used to signal the defining feature-sets of simple-tasks. When a large number of simple-tasks and features are involved, however, it is convenient to lay out the values of different simple-tasks on the different features in a matrix. (I have abbreviated the feature names for typographical convenience.)

Inspecting the matrix of tasks and features, it is possible to notice ex-

Table 5.1: Simple-task dictionary for Minimalist's TAG of TEDIT

Simple-Tasks	T.ins?	T.Del?	T.Unit	S.Exist?	S.Unit	S.Rem?	B/A
				Features			
insert before char	yes	no	char	no	N/A	N/A	before
insert after char	yes	no	char	no	N/A	N/A	after
insert before word	yes	no	word	no	N/A	N/A	before
insert after word	yes	no	word	no	N/A	N/A	after
delete character	no	yes	char	no	N/A	N/A	any
delete word	no	yes	word	no	N/A	N/A	any
delete string	no	yes	string	no	N/A	N/A	any
replace string	yes	yes	string	no	N/A	N/A	any
copy char before char	yes	no	char	yes	char	no	before
copy char after char	yes	no	char	yes	char	no	after
copy char before word	yes	no	word	yes	char	no	before
copy char after word	yes	no	word	yes	char	no	after
copy word before char	yes	no	char	yes	word	no	before
copy word after char	yes	no	char	yes	word	no	after
copy word before word	yes	no	word	yes	word	no	before
copy word after word	yes	no	word	yes	word	no	after
copy string before char	yes	no	char	yes	string	no	before
copy string after char	yes	no	char	yes	string	no	after
copy string before word	yes	no	word	yes	string	no	before
copy string after word	yes	no	word	yes	string	no	after
move char before char	yes	no	char	yes	char	yes	before
move char after char	yes	no	char	yes	char	yes	after
move char before word	yes	no	word	yes	char	yes	before
move char after word	yes	no	word	yes	char	yes	after
move word before char	yes	no	char	yes	word	yes	before
move word after char	yes	no	char	yes	word	yes	after
move word before word	yes	no	word	yes	word	yes	before
move word after word	yes	no	word	yes	word	yes	after
move string before char	yes	no	char	yes	string	yes	before
move string after char	yes	no	char	yes	string	yes	after
move string before word	yes	no	word	yes	string	yes	before
move string after word	yes	no	word	yes	string	yes	after
copy-replace string with char	yes	yes	string	yes	char	no	N/A
copy-replace string with word	yes	yes	string	yes	word	no	N/A
copy-replace string with string	yes	yes	string	yes	string	no	N/A
copy-replace string with char	yes	yes	string	yes	char	no	N/A
move-replace string with word	yes	yes	string	yes	word	no	N/A
move-replace string with string	yes	yes	string	yes	string	no	N/A

amples of "semantic incompleteness" (Payne and Green, 1986). This is the kind of semantic inconsistency which occurs when operations which are available with some parameters are not available with others, for no reason that is apparent to the user. In this case, it seems peculiar that it is possible to *delete* and to *copy* characters words or strings, but only possible to *replace* strings, and only possible to *insert* by characters or words. These semantic inconsistencies are highlighted in the feature matrix (and could be found automatically from such a matrix), because the changing Target.Unit feature defines the groups of related tasks, but it does not take all its possible values. Semantic incompleteness has repercussions on the mapping of tasks to actions, as we shall see.

Mapping tasks onto actions First, let us assume that users can perceive a common structure for all the commands, in which the target is marked, and then some subsequent actions performed. In TAG this common structure might be represented by a single higher-order rule:

M1 Task [F()] → mark [Target.Del?, Target.Unit,
        Before/After]
   + do [F()]

(Rules in this task action grammar are labelled Mi for "Minimalist's rule i". Notational remark: The F() symbol is merely a convenient typographical short-hand, denoting that *all* the features of the task, and of the "do" sub-task are liable to influence their subsequent rewriting. The names "mark" and "do" for the non-terminals are just arbitrary labels: they have no semantic status in the grammar.)

According to rule M1 then, the initial marking of a target depends on the text unit involved, and whether or not it is to be subsequently deleted. This in many ways seems the most natural semantic organisation, because of the way extending selection of a text string automatically enforces subsequent deletion of it, as suggested by the term "extend pending delete".

The mark subtask can in turn be rewritten as follows:

M2 mark [T.Unit, B/A] → point [B/A] + click [T.Unit]
M3 point [B/A = before] → "point to left-half of unit"
M4 point [B/A = after] → "point to right-half of unit"
M5 point [B/A = any] → "point to unit"
M6 click [T.Unit=char] → "click left button"
M7 click [T.Unit=word] → "click middle button"
M8 mark [T.Del?=yes] → mark [T.Unit=value-from-goal]
        + "point"
        + "click right button"

Rules M3 and M4 describe the way that the position of subsequent actions depends on the precise location of the pointing, and capture the generality in this operation between characters and words. However, they do not represent the "natural" aspect of this design whereby the point-of-action always moves to the *nearest* side of the pointed-to-unit. Such aspects can be captured in TAG, by describing a featural semantics in the action domain as well as in the task domain. Payne and Green (1986, section 2.3) demonstrate this facility for lexical action specifications. In the current case, the before/after semantics seem too simple to warrant such a diversion from our main enterprise.

Rule M8 illustrates two further aspects of TAG. Firstly, we see the standard recursive capacity to rewrite one non-terminal symbol in terms of itself. Secondly, we see a feature taking a value "from the goal". The idea here is that some tasks are performed in related ways, but differing according to the details of the user's current goal. In this case, one can extend an initial marking whether it is of a character, or a word. The value of such features is assumed to be passed to the task-action grammar as a parameter from some higher-level planning system (see Payne and Green, 1986).

A problem with rules M1 to M8 is that extend-pending-delete is not the only way to delete portions of text, it is also possible to delete words or characters by marking them without extending and then pressing "DEL", and to delete characters words or strings by depressing the control key during selection, and then releasing. So if users want to remember and use these redundant deletion facilities, some restructuring of the grammar is required. We can speculate that this may cause some learning problems if users have first learnt to use the system without these redundant facilities. As we shall see below, however, the new structuring is vital if all the subtle consistency of the task language is to be captured.

Let us first consider the DEL key option. If this is to be used to delete selected characters and words, M1 is an inaccurate rule: the marking method need not depend on the value of T.Del. (In this and subsequent examples letters are appended to rule numbers to denote new generations of rules that are needed as our analysis develops.)

> M1a Task [F()] → mark [T.Unit, B/A]
>   + do [F()]

Rule M8 must now be rewritten as follows (rules M2 through M7 can remain unaltered):

> M8a mark [T.Unit=string] → mark [T.Unit=value-from-goal]
>   + "point"
>   + "click right button"

The CTRL option is harder to accommodate by small alterations to our existing rules. Yet its tidy accomodation is vital, because the control key works consistently on target and source selections, as we will discuss below.

Because the control key has exactly the same effect on all three kinds of marking (that is, of characters, words and strings), it is most convenient to include it as an option on a single generalised token for marking:

```
M1b Task [F()] → genmark [T.Unit, T.Del?, B/A]
                 + do [F()]
M1.1 genmark [T.Ins?=no, T.Unit, T.Del?, B/A] →
                 key [T.Del?]
                 + mark [T.Unit, B/A]
                 + release [T.Del?]
M1.1' genmark [T.Unit, T.Del?, B/A] → mark [T.Unit, B/A]
M9 key [T.Del?=yes] → "depress CTRL"
M10 key [T.Del?=no] → NULL
M11 release [T.Del?=yes] → "release CTRL"
M12 release [T.Del?=no] → NULL
```

Rules M1.1 and M1.1' are new rules which must sit between the new rule M1b and the rules M2 to M8. There are two alternative rewritings because of the redundancy of CTRL-delete operation – text can happily be deleted without using the CTRL option. These alternative rules exemplify a general need to produce over-large acceptance grammars when the desire is to analyse all possible structurings of a task, rather than using maximally economic generative grammars to describe a minimal, sufficient set of task-action mappings. It is clear that representation of alternative methods will in general require more rules in a task-action grammar, reflecting, quite appropriately, extra learning effort.

There is one more point worth noting about the grammar for the subset of the task language developed so far. Marking a target while depressing the control key enforces immediate deletion of the target, excluding all the replace operations which demand a pending deletion – this has been represented in the grammar by forcing T.Ins?=no in the rewriting of the CTRL deletion option (rule M1.1). This device has repercussions for the representation of CTRL-deletion in source selection, as we shall see. (Replacement can of course be achieved using CTRL selection, but only by composing separate deletion and insertion tasks – the visible removal of the target will happen before the insert, move or copy operations are executed.)

The *do* subtask has three main branches, inserting, deleting with the delete key, and copying or moving text in the various forms of those operations.

The insertion of text is straightforwardly defined:

```
M13 do [T.Ins?=yes, S.Exist?=no] → TYPE
```

This rewrite rule completes both the insertion and the replacement simple-tasks. But note that at this stage, if the rules M1a or M1b to M8a are used, we have specified the actions for inserting text without any reference to the T.Del? feature. Yet that feature is all that distinguishes Insertion from Replacement – or rather it would be were it not for the tying of the string unit to the replace task (the semantic incompleteness we noted above). What we are seeing here is a repercussion of that semantic incompleteness on the structuring of the task-action mappings. According to this TAG analysis users need to distinguish replacement from insertion by explicitly associating replacement with the unit string. The theory of task-action grammars is not yet well enough developed to make explicit empirical predictions concerning the problems that may result from this need to represent certain structural aspects of a language in terms of intuitively low-salience featural distinctions. Nor have any metrics over the grammar been put forward to allow this issue to be detected unintelligently. Nevertheless, it is hoped that this example illustrates another aspect of TAG's potential as a tool for analysis.

Deletion using the delete key is once more straightforward:

M14 do [T.Del?=yes, S.Exist?=no] → "DEL"

The moving and copying operations are more complex, and for this task language more problematic. For these reasons they are the best illustrators of the power of TAG. All the copy and move tasks are marked by the semantic component S.Exist=yes. They are all accomplished by depressing the SHIFT key and then selecting a source in much the same way as the target is selected, in the *mark* subtask. All the *do* subtasks for move and copy can therefore be rewritten by a single rule schema:

```
M15 do [S.Exist=yes, S.Rem?, S.Unit] → "depress SHIFT"
   + genmark [              T.Unit=S.Unit,
                            T.Del?=S.Rem?
                            B/A=any]
                            + "release SHIFT"
```

This rule-schema illustrates the power of TAG in capturing the consistent design of the TEDIT task language. In fact it utilises a device not seen in previous papers, although not one that requires any extension of the formal apparatus. Allowing one feature to pass its value to another, which in turn determines the rewriting of a consituent in another portion of the grammar (T.Unit=S.Unit, T.Del?=S.Rem?) is a powerful device which allows a single rule schema to capture a subtle consistency.

Unfortunately rule M15 does not quite work! The *only* way to remove the source (for the move commands) is by using CTRL-deletion, so rule M1.1 *must* be used to rewrite the genmark component of the secondary selection. But, because of the alternatives for deleting targets, we have an alternative rule M1.1', with no way of determining which must fire, except that we have specified that rule M1.1 should only apply when we are not replacing the target – something that is no longer a valid restriction when we consider secondary selection. Yet if we remove that condition:

M1.1a genmark [T.Unit, B/A, T.Del?]→ key [T.Del?]
+ mark [T.Unit, B/A]
+ release [T.Del?]

we are left with two alternative rules (M1.1a and M1.1') with no distinguishing features. This is not a problem in general, but it is a problem when we need to enforce choice of one particular alternative in some circumstances, as we do here.

Far from being disappointed at the difficulty of expressing the details of CTRL-deletion in TAG, I am encouraged to believe that our difficulty is illustrative of several psychological consequences of TEDIT's design. Let me speculate.

Novice users with minimal device spaces may well make exactly the error that rules M1.1a and M1.1' allow, namely failing to depress CTRL when the secondary selection is a string. I have no evidence to support this prediction, but the error derives from quite a straightforward over-generalisation: "string selection always leads to deletion". (This problem arises, of course, from the semantic incompleteness we noted above.) To overcome this difficulty within the minimal device space, users must develop some way of choosing between M1.1a and M1.1'. I predict that some users will *never* use CTRL-deletion on target selections, whereas some will *always* use it, forsaking the other modes of deletion, and the replacement operations. The latter strategy is readily described in TAG, by simply deleting rules M1.1' and M14, but it is clearly sub-optimal. The former strategy and intermediate ones in which CTRL-deletion is used in some situations but not in others, can also be represented in TAG rules of course, but only by specifying extra features in existing rules, or completely new rules.

So the design of TEDIT seems to be spoilt by the inconsistency of the CTRL-delete option, at least for users who have only constructed the minimal device space.

5.3.2   Elaborator's TAG for TEDIT

If the minimal device space were elaborated just enough to include the cursor as an independent conceptual entity, as we might expect with a user who comes to TEDIT from a keyboard-driven screen-editor, would structuring the task-action mappings become any easier? In fact, the answer is no – just the opposite, the consistent structuring of the task-action mappings becomes very hard to represent.

If the cursor is treated as a device space object, then cursor-movements should be treated as simple-tasks. But if cursor movements (to before or after words or characters) are separate simple-tasks, then source-selections, as in the move and copy commands, cannot be constructed as generalisatons of the "mark" operation, as they were above, for that has become tagged as a cursor movement. Secondary, or source selections do *not* move the cursor. The most simple conception of the cursor, therefore, will not allow a user to capitalise on the organisational consistency of the task language. Furthermore, if initial "marking" is construed simply as a cursor movement, we might expect the mark-DEL and CTRL-deletion methods to be ignored, leaving back-deletions (not considered in the TAG analysis) and extend-pending-delete as the favoured methods for deletion.

It is possible, nevertheless, to elaborate the minimal device space, so as to give previously syntactic-only marking operations a figurative semantics. If the user constructs the concept of a "selection", which embodies a richer construction of the cursor concept, and which can be either a target or a source in the goal space then many sub-operations become operators on the device space, and hence simple-tasks. For example, the various selections of targets and sources become simple-tasks in their own right, as do the subsequent actions on those selections. Naturally, the "action" simple-tasks will have strict *preconditions* – usually they will depend on a particular type of selection already having been made – but this notion of preconditions is entirely consonant with our view of simple-tasks as problem space operators. Indeed, preconditions and postconditions are the major way of defining operators in the problem solving literature, and it is inevitable that they will have a considerable role in the definition of simple-tasks in task-action grammar analyses, although they have not done in many of the presented analyses to date; see however Payne and Green's (1986) analysis of Moran's (1981) "EG" message system. Following that treatment, in the current example we represent the vital preconditions as additional features in the simple-task dictionary.

Particularly important, in view of the difficulty with the CTRL-deletion options experienced in the Minimalist's TAG, is the possiblity of having "selection of a target for deletion only" as a separate simple-task, along with the alternative "selection of a target for deletion or insertion". Two of the corresponding "action" tasks will be "delete a target that has been selected specially for deletion" and "delete any selected target". We will see below how the specification of these separate tasks allows us to define rule schemas which overcome the problems we experienced above.

The new set of simple-tasks will be: select a character or word as target, for any subsequent action or for deletion only; select a string as target for replacement or deletion; select a character, word or string as source for copying or moving; insert text; delete target selected for deletion; delete target selected for anything; copy selected-source; and move selected-source.

The semantic features which organise the set of simple-tasks will also be different in the elaborated device space. I suggest that the following features offer a natural organisation of the task world:

Feature	Possible Values
Selection/Action	selection, action
Source/Target	source, target
Unit	character, word, string
Ins?	yes, no, any
Del?	yes, no, any
Context	no selection, target selected for deletion, target selected for anything, source selected for copy, source selected for move
Before/After	before, after, any

Once again there are cooccurrence restrictions. The Unit feature only applies to selections, not actions. The Context feature represents the differing preconditions for the actions. It turns out that all the "action" simple-tasks have very demanding preconditions, but not quite unique ones, so it is still necessary to define them in terms of other features in order to expose the similarities and differences between them.

Table 5.2 shows the featural makeup of all the simple-tasks according to these features.

Mapping tasks onto actions According to the feature matrix the user perceives two main types of task, *selections* and *actions*. These task types must be rewritten separately.

First let us consider selections:

> E1 Task [Selection/Action=selection, Unit, Source/Target,B/A]
> $\rightarrow$ key [Source/Target]
> + key1 [Del?]
> + mark [Unit, B/A]

(Rules in this task-action grammar are labelled Ei for "Elaborator's rule i".)

> E2 key [Source/Target=source] $\rightarrow$ "depress SHIFT"
> E3 key [Source/Target=target] $\rightarrow$ NULL
> E4 key1 [Del?=yes] $\rightarrow$ "depress CTRL"
> E5 key1 [Del?=no] $\rightarrow$ NULL
> E6 mark [Unit, B/A] $\rightarrow$ point [B/A]
> + click [Unit]
> E7 point [B/A= before] $\rightarrow$ "point to left half of unit"
> E8 point [B/A = after] $\rightarrow$ "point to right half of unit"
> E9 point [B/A = any] $\rightarrow$ "point to unit"
> E10 click [Unit=char] $\rightarrow$ "click left button"

Table 5.2: Simple-Task dictionary for Elaborator's TAG of TEDIT

Simple-Tasks	Features						
	Selection/Action	Source/Target	Unit	Ins?	Del?	Context	B/A
select a char as target for anything with cursor before	selection	target	char	any	any	no selection	before
select a char as target for anything with cursor after	selection	target	char	any	any	no selection	after
select a word as target for anything with cursor before	selection	target	word	any	any	no selection	before
select a word as target for anything with cursor after	selection	target	word	any	any	no selection	after
select a char as target for deletion only	selection	target	char	no	yes	no selection	any
select a word as target for deletion only	selection	target	word	no	yes	no selection	any
select a string as target for replacement	selection	target	string	yes	yes	no selection	any
select a string as target for deletion only	selection	target	string	no	yes	no selection	any
select a char as source for copying	selection	source	char	yes	no	no selection	N/A
select a word as source for copying	selection	source	word	yes	no	no selection	N/A
select a string as source for copying	selection	source	string	yes	no	no selection	N/A
select a char as source for moving	selection	source	char	yes	yes	no selection	N/A
select a word as source for moving	selection	source	string	yes	yes	no selection	N/A
select a string as source for moving	selection	source	string	yes	yes	no selection	N/A
insert text	action	target	N/A	yes	any	target for anything	N/A
delete target selected for deletion	action	target	N/A	no	yes	target for deletion	N/A
delete target selected for anything	action	target	N/A	no	yes	target for anything	N/A
copy selected-source	action	source	N/A	yes	any	source for copy	N/A
move selected-source	action	source	N/A	yes	yes	source for move	N/A

E11 click [Unit=word] → "click middle button"
E12 mark [Unit=string] → mark [Unit=value-from-goal]
+ "point and click right button"

Rules E1 to E12 completely define all the selection simple-tasks. They capture the consistency of the marking operation in both target and source selection, and the effect of the control key on both operations. Many of the rules will look familiar for example, the mark operation is the same as in the Minimalist's TAG.

Indeed, these rules suffer the same problems as the Minimalist's TAG. In particular, the inconsistent status of CTRL-deletion is not tackled by these rules. Rule E4 dictates that CTRL must always be depressed if the target is to be deleted, but as we have seen, this is only true of secondary (source) selections.

However, with our elaborated device space it is possible to construct rule schemas which overcome this problem. This capacity is due to the deconstruction of sentential operations into components making it meaningful to specify whether the current selection operation applies to a target or a source.

E1a Task [Selection/Action=selection, Unit, Ins?, Del?,
Source/Target, B/A]→
key [Source/Target]
+ key1 [Del?, Source/Target]
+ mark [Unit, B/A]
E4a key1 [Del?=yes, Source/Target = source] →
"depress CTRL"
E4.1 key1 [Del?=yes, Ins?=no, Source/Target = target] →
"depress CTRL"
E4.1′ key1 [Del?=any, Source/Target = target] → NULL

Rule E4a now specifies that CTRL must be depressed during source selections for deletion, but rules E4.1 and E4.1′ express the fact that CTRL is a possible but not essential way of deleting target selections, and that it cannot be used for the replacement operation. (In contrast with the Minimalist's TAG it is now valid to force Ins?=no.) The conflict that existed in the Minimalist's task-action grammar is resolved by these rules for rewriting operators on the elaborated device space.

The "action" simple-tasks complete the elaborator's task-action grammar:

E13 Task [Selection/Action=action, Source/Target=target,
Ins?=yes, Del?=any, Context=target selected for
anything]
→ TYPE
E14 Task [Selection/Action=action, Source/Target=target,
Ins?=no, Del?=yes, Context=target selected for
anything]

→ "DEL"
E15 Task [Selection/Action=action, Source/Target=target
        Context=target selected for deletion]
            → "release CTRL"
E16 Task [Selection/Action=action, Source/Target=source,
        Context=source selected for copy]
            → "release SHIFT"
E17 Task [Selection/Action=action, Source/Target=source,
        Context=source selected for move]
            → "release SHIFT and release CTRL"

It may appear unduly clumsy to use separate top-level rules for each of the five simple-tasks. This seems unavoidable for E13 and E14, but could not the commonality between E15, E16 and E17 be exploited, perhaps by using E15 and E16 in the definition of E17. In fact this is impossible to do in TAG, and inappropriate, for the following reason. When a secondary source has been selected for moving, by pressing both SHIFT and CTRL, nothing happens until *both* these keys are released. If just one of the keys is released the user is completely trapped in a mode, unable to do anything until the other is released to complete the operation. So "release SHIFT and release CTRL" is indeed a single operation, as shown in rule E17.

Rules E1 to E17 describe the complete set of mappings from tasks to actions for the subset of TEDIT under consideration. The Elaborator's task-action grammar does not suffer from the representational problems of the Minimalist's grammar, in which, remember, it was very difficult to achieve an accurate mapping without forsaking some of TEDIT's facilities. For TEDIT, it seems, the elaboration of a device space to include the concepts of target and source selections as autonomous entities makes the full structure of the task language easier to learn.

A more detailed comparison of the two task-action grammars is provided in the next subsection. The general insights the twin TAG analyses have afforded are discussed in the final section.

5.3.3   A comparison of the two TAGs

The summary statistics for the final generation of each task-action grammar are as follows:

The Minimalist's TAG has 38 simple-tasks described by 7 semantic features. The simple-tasks are rewritten into action specifications by 17 rule-schemas. It must be remembered, however, that these 17 rule-schemas only achieve an imperfect mapping, providing no way of distinguishing which of two alternative rules must be used for deletion of source-selections.

The Elaborator's TAG has 19 simple-tasks described by 7 semantic features. The simple-tasks are rewritten into action specifications by 19 rule-schemas. Unlike the Minimalist's TAG the Elaborator's TAG provides a complete and correct representation of all alternative task-action mappings.

One must be careful in drawing conclusions from statistical comparisons between the two grammars. We will discuss each of the major comparisons in turn. Firstly, the Elaborator's TAG has many fewer simple-tasks than the Minimalist's TAG. This relationship is not neccessarily a uniform one between minimal and elaborated device spaces. By elaborating a device space, the learner treats many of the components of simple-tasks in the minimal device space as autonomous simple-tasks. Previously "sentential" operations are deconstructed into their constituent parts. In this instance this results in a lower total number, because many of the minimalist's simple-tasks share components.

In contrast we believe the fact that the same number of features (some taking a larger number of distinct values) is needed to describe fewer simple-tasks does reflect a general idea. Elaborated device spaces allow a semantically richer representation of operators, and in the TAG notation this can only be reflected by the features.

Finally, if the elaborate device space is advantageous to users, why does it lead to a grammar with so many top-level rules? This is especially mysterious given the special status which has previously been given to top-level rules as complexity metrics.

There are several layers of answer to this question. First, the top-level rule complexity metric is designed to give crude comparisons between the learnability of different but similar task languages; our aim in offering alternative task-action grammars for the same language is to explore TAG's potential for delivering insights into the configuration of the design. When TAG is being used in this way numeric metrics can become meaningless, as remarked by Payne and Green (1986, section 2.4). If the goals had been different, and we had sought to make a broad-brush comparison between TEDIT and an alternative design, then we would have attempted to ensure that both the to-be-compared TAG descriptions made appropriate and equivalent assumptions about the conceptual models of the "ideal learner". The best way to tackle this problem depends on the particular purpose of the comparison, but the approach that has been adopted in previous descriptions of TAG is to develop the most economical task-action grammar, to give a lower-bound on perceived complexity.

Related to this issue, it is worth noting that the single top-level rule in the Minimalist's TAG does not do much expressive work, it highlights a very weak structural uniformity among the commands. There is an important lesson here: it is often possible to invent weak top-level rules of this kind, but if systems are to be compared accurately it is necessary to adopt a consistent "style" in order to render the rule-count metrics meaningful. The ideal solution to this problem would of course be an algorithm to generate task-action grammars of task-languages automatically. More work is needed here.

Thirdly, it is possible that the Elaborator's TAG really would take more learning than the Minimalist's version. However, once it has been learned it allows a more complete description of the language's structure, and more flexible problem solving. The Minimalist's TEDIT tempts the user into all sorts of self-imposed mode traps (as do all operational accounts of operators). For example, according to the Minimalist's TAG, when the user has set out to copy a block of text by marking a target character, he or she has no choice but to carry on with the operation – simple-tasks are not interruptible. Users with elaborated device spaces offering figurative accounts of operators face many fewer such problems.

With this last point, our discussion of the quantitative differences between the two task-action grammars for TEDIT has lead us into a discussion of qualtitative differences between the two models. Such qualitative issues are the major focus of this chapter. In the final section we recap the inferences about the learning of TEDIT that the TAG analyses have afforded, and consider some of their practical and theoretical implications.

5.4   Conclusions

The main goal of this chapter was to explore the potential of task-action grammars for reasoning about the learning demands of real-world, current-technology user interfaces. This exploration has been limited to the analysis of a variety of possibly perceived structures in a single example task language, and the drawing out some implications of these structures for learning.

Beyond this, and beyond previous papers, the TAG notation has been integrated with the yoked state space hypothesis to explore the ways in which alternative conceptual models of the device may influence the perceived structure of task languages.

The success or failure of this enterprise must be judged on the quality, and the implications, of the insights into learning difficulties gained from the TAG analyses.

5.4.1   Summary

The Minimalist's TAG of TEDIT highlighted the following difficulties, together with some speculated behavioural coping strategies that may be adopted:

i The provision of redundant deletion methods may force learners that have previously used just one method to restructure a large part of their initial task-action grammar. If users neglect to tackle this cognitive load, then the learning of move and copy operations may be made much more difficult, as it becomes difficult to capitalise on the consistent structure of target and source selections.

ii It seems necessary to distinguish the tasks of insertion and replacement in terms of the text-units to which they apply. Only strings can be replaced; text can be inserted by characters or words. We might speculate that learners will confuse insertion and replacement tasks. Because of the more elaborate action sequence needed to mark strings for replacement learners are unlikely to accidentally replace strings adjacent to which they intended to insert text, but they might sometimes insert text next to words they had intended to replace.

iii Although target deletions can be achieved by first marking a string, or by typing "DEL" after marking, source deletions can only be achieved by depressing "CTRL" while marking. This makes it difficult for learners to capture the aspects of target and souce operations which are consistent, without adopting an erroneous conception of source deletion methods. (See the discussion of rules M1.1a and M1.1' in section 5.3.1.) This may result in users committing an error of ommission: failing to use "CTRL" when the source selection is a string. It may alternatively result in users adopting the coping strategy of always or never using CTRL-deletion for the target which will result in a cost of redundant keystrokes for some deletion tasks.

One criticism of the Minimalist's TAG may be that it makes the unrealistic assumption that the cursor is not an autonomous conceptual entity in the learner's device space. But it is not clear that this assumption is unrealistic for first-time text editor users who have no experience with a keyboard-controlled cursor or with a typewriter, for such users might happily cope with a concept of a mouse-position marker, or pointer, and ignore the cursor (as discussed in section 5.2.2). Furthermore, including a simple cursor-concept in the device space actually makes the task language much harder to learn, for it inhibits the possibility of representing the common structure of target and source operations (as discussed in section 5.3.2).

The Elaborator's TAG describes perceived structures for users with an elaborated device space, in which the notions of target and source selections have the status of autonomous conceptual entities. This TAG still suffers from difficulties **i** and **ii** which both result directly from the inherent semantic inconsistency of the TEDIT language. However, Elaborator's TAG does overcome difficulty **iii**: learners with an elaborated device space are able to represent the consistent elements of target and source selection while still maintaining a sufficiently differentiated model for how to perform deletion of these separate selections.

In a nutshell, the two main points emerging from the TAG analyses of TEDIT are:

Firstly, TEDIT's task language presents some awkward structures for beginners to learn, and these awkwardnesses may lead to errors and inefficient strategies;

Secondly, users with an elaborated conceptual model of the device will be better able to represent the consistent structure of TEDIT than will users with "minimal" device spaces, which fail to give a semantic interpretation to some of TEDIT's primitive operations.

5.4.2   Discussion

The advantage of the Elaborator's over the Minimalist's task-action grammar is an example of the potential influence of conceptual models on the mental representation of procedural skills. Theories of learning in cognitive psychology (eg Anderson, 1982, 1987; van Lehn 1983) often assume that the development of procedural skills can be modelled by mechanical changes to algorithmic methods; this view denies the important role played by "understanding" in its various forms. The yoked state space hypothesis explicity addresses one aspect of understanding, and the current integration of this with task-action grammars shows one line of influence between understanding and method.

The cognitive advantages of the Elaborator's TAG carry an important message for instruction. It may be wise to explicitly instruct novices with the selection concept, rather than simply teaching them how to perform tasks and running the risk of them forming purely operational accounts of all the methods.

In the absence of any such instruction, it is tempting to speculate that TEDIT is better suited as a text editor for the computer-experienced than for the raw novice – and, indeed, the former is the obvious target population for an editor that runs on Lisp machines.

Together with our observations about the difficulties of learning TEDIT's task language, and the advantages of an appropriate conceptual model, we have speculated about some of the specific behavioural consequences, in terms of errors, or inefficient strategies, that may result from weak conceptual models, or from inappropriately perceived task language structures. It is clearly a shortcoming of the current chapter that these predictions have not been put to the empirical test. Nevertheless, the speculations about inefficient behaviour suggest an exciting avenue for exploration: the use of task-action grammar and yoked state space analyses in the provision of intelligent on-line advice to novice users.

To exemplify in terms of the above summary, we have observed that users with minimal device spaces, may utilise only some of the available methods for deleting text. This can certainly lead to inefficient usage: Users who only use the extend-pending-delete method will perform redundant, fiddly, pointing; sole reliance on CTRL-deletion methd could lead users to remark current selections. None of these is a major problem for the user, but they are illustrative of the "acquired mediocrity" that is so often observed, even in experienced users. Furthermore, all these behaviours can be *automatically* noticed, and diagnosed in terms of the TAG model presented above. This diagnosis supports the prospect of remedial coaching that goes beyond mere demonstration of the "correct" method. Although it has recently become fashionable to question whether such diagnosis-based coaching may be any more advantageous than simple reiteration of correct procedures (eg Sleeman, 1987) it should be noted that current diagnoses are typically limited to mal-rules or bugs which assume purely procedural "skill-as-method" performance models (Payne, 1987a). The analyses suggested here push beyond this, towards diagnosis of learners' conceptual models.

The instructional benefits of conceptual-model diagnoses are yet to be explored. This potential use of TAG, as the basis of a student model for intelligent computer-assisted instruction, is the current focus for development of the ideas presented here.

Acknowledgements

Tom Moran suggested TEDIT as a good testing ground for TAG, and worked on some analyses of its structure himself. The analysis presented here has borrowed from Tom's insights, but differs quite substantially. Discussions with Tom were also instrumental in the genesis of the goal space and device space ideas, which are inspired by his ETIT analysis (Moran, 1983). The other major influence on this work is Thomas Green, who supervised my initial work on TAG. Finally, I owe a large debt to the editors of this book, who covered an earlier draft with red ink, forcing me to express my ideas more clearly.

This research is supported by Alvey/SERC under grant GR/D 60355.

References

Anderson, J.R. (1982) Acquisition of cognitive skill. *Psychological Review*, **89**, 369-406.
Anderson, J.R. (1983) *The Architecture of Cognition*. Cambridge, Mass.: Harvard University Press.

Anderson, J.R. (1987) Skill-acquisition: Compilation of weak-method problem solutions. *Psychological Review*, **94**, 192-210.

Barnard, P.J., Hammond, N.V., Morton, J. Long, J. and Clark, I.A. (1981) Consistency and compatibility in human- computer dialogue. *International Journal of Man-Machine Studies*, **15**, 87-134.

Carroll, J.M. (1982) Learning, using, and designing command paradigms. *Human Learning*, **1**, 31-62.

DiSessa, A. (1986) Models of computation. In D.A. Norman and S.W. Draper (Eds) *User-Centred System Design*. Hillsdale, NJ: Erlbaum.

Douglas, S. (1983) Learning to text edit: Semantics in procedural skill acquisition. Unpublished doctoral dissertation, Stanford University, Palo Alto, California.

Green, T.R.G. (1980) Programming as a cognitive activity. In H.T. Smith and T.R.G. Green (Eds.) *Human Interaction with Computers*. London: Academic Press.

Green, T.R.G. and Payne, S.J. (1984) Organisation and learnability in computer languages. *International Journal of Man-Machine Studies*, **21**, 7-18.

Greeno, J. (1983) Conceptual entities. In D.Gentner and A. Stevens (Eds) *Mental Models*. Hillsdale, NJ: Erlbaum.

Laird, J.E. Newell, A. and Rosenbloom, P.S. (1987) SOAR: An architecture for general intelligence. *Artificial Intelligence*, **33**, 1-64.

Long, J. This volume.

Moran, T.P. (1981) Command language grammar: A representation for the user interface of interactive computer systems. *International Journal of Man-Machine Studies*, **15**, 3-50.

Moran, T.P. (1983) Getting into a system. External-internal task mapping analysis. Proceedings of the CHI 83 Conference on Human Factors in Computing Systems, 45-49, New York: ACM.

Newell, A. (1973a) Production systems. In W.G. Chase (Ed) *Visual Information Processing*. New York: Academic Press.

Newell, A. (1973b) You can't play twenty questions with nature and win. In W.G. Chase (Ed) *Visual Information Processing*. New York: Academic Press.

Newell, A. and Simon, H.A. (1972) *Human Problem Solving*. Englewood Cliffs, NJ: Prentice-Hall.

Payne, S.J. (1985) Task-action grammars. In B. Shackel (Ed.) Interact 84 (Conference Proceedings). North Holland.

Payne, S.J. (1987a) Methods and mental models in theories of cognitive skill. In J. Self (Ed.) *Artificial Intelligence and Human Learning: Intelligent Computer-Aided Instruction*. London: Chapman & Hall.

Payne, S.J. (1987b) Complex problem spaces. Modelling the knowledge needed to use interactive devices. Proc. Interact 87.

Payne, S.J. and Green, T.R.G. (1986) Task-action grammars: A model of the mental representation of task languages. *Human-Computer Interaction*, **2**, 93-133.

Payne, S.J. and Green, T.R.G. (in press) The structure of command languages: an experiment on task-action grammar. *International Journal of Man-Machine Studies*.

Reisner, P (1981) Formal grammar and the design of an interactive system. *IEEE Transactions on Software Engineering*, **5**, 229-240.

Rumelhart, D.E. and McClelland, J.L. (1986) *Parallel Distributed Processing. Explorations in the Micro-structure of Cognition*. Cambridge, Mass: MIT Press.

Scapin, D.L. (1982) Generation effect, structuring and computer commands. *Behaviour and Information Technology*, **1**, 401-410.

Sleeman, D. (1987) Some challenges for intelligent tutoring systems. Proc. IJCAI 87, 1166-1168.

van Lehn, K. (1983) *Feclicity conditions for human skill acquisition: Validating an AI-Based theory.* Xerox Palo Alto Research Center. Technical Report CIS-21.

6

Expressing research findings to have a practical influence on design

Paul Buckley

6.1   Introduction

The ultimate aim of applied research in Human-Computer Interaction (HCI) is to improve the design of information technology devices. Where design and research are conducted separately, any such improvement can be made only through the research findings being taken up and applied by those responsible for design. In some domains there might exist an authority responsible for assessing research findings and passing the results on to designers. Such a scheme is operated by the British Standards Institute, and by various professional organisations such as the Civil Aviation Authority. These bodies typically embody design criteria in legislation, or through a mechanism of approval, so constraining acceptable design solutions.

Currently there is no comparable body assessing standards of human-computer interface technology. This may be partly because the idea of having standards in such a wide and unexplored field is premature and controversial, and also because there is little demand for standards. Whereas the risks inherent in the use of inflammable foam in furniture are obvious and potentially serious, it is difficult to be so unequivocal about the problems of, say, 10% decrements in performance time as a result of an unsuitable command language.

Given that no standards body exists with respect to quality of human interfaces, research findings must be communicated directly to designers in order that design decisions are informed. There is no formal route for this communication and so concerned researchers may have to set up the communication channels themselves. In establishing the channels, account will need to be taken of both the requirements for the physical system of distribution, such as the appropriate vehicle for the material – journal, report or magazine, etc.; and of the form of expression of findings that will be acceptable and useful, such as for example, technical data, or as a general model of behaviour.

In the terms of the framework for HCI work (Long, Chapter 1, this volume), the expression of the research findings in a suitable way for designers will encourage and aid 'synthesis' of the information into the 'real world of work'. There is an assumption here that it is the responsibility of the designers, not the researchers, to actually carry out the synthesis phase. Other authors share this assumption. Probably the most influential of these are Card, Moran and Newell (1983). They believe that the designer should be the primary agent in applying psychology to systems. For this application to be aided and encouraged:

> '...it is necessary that a psychology of interface design be cast in terms homogeneous with those commonly used in other parts of computer science and that it be packaged in handbooks that make its application easy. Thus, the system designer in our scenario finds the design handbook more efficient than plunging blindly into code with his Pascal compiler, although he may still find it profitable to engage in exploratory implementation.' (p13)

In terms of the framework, the designer is aided by an appropriate applications representation. It is important to cast this representation in an accessible and useful form. Indeed, the commercial exploitation of the results may rest on the appropriate expression. In the extract above, the design handbook is assumed to provide a satisfactory representation of research to designers. This assumption is based first on the scenario of design practice illustrated above, and second, on a principle that information materials should be provided in homogeneous terms. With respect to the first point, it is unclear whether this scenario of handbook use in design is representative in the field of system design, or applicable to particular design domains. Some designers might simply not use design handbooks, or even know of their existence, but instead subscribe to a wide range of human factors journals. With respect to the principle of homogeneity, we also assume that the most influential expressions of design information are either those used in current design practice, or similar to information currently used. We prefer to describe our aim as the identification of compatible rather than homogeneous expressions, since the latter implies a degree of similarity that unnecessarily constrains the range of appropriate applications representations. However, both the homogeneity principle and the compatibility approach assume that designers will apply human factors information if it is presented in a familar form. The acceptability of this assumption will be examined later, in the context of videotex dialogue design.

Galer and Russell (1987) contrast three presentation modes of information that may be offered to designers. Paper modes include manuals, checklists, methodologies and procedures. Computer modes include computer aided instruction and decision support systems. People modes include seminars, audio-visual presentations and training. An applications representation may be characterised both by mode and by the sort of information presented in that mode. To clarify this distinction, we will describe applications representations as consisting of a 'vehicle' and a 'form'. Vehicles are the carriers of representations, and correspond to the classes of presentation mode. Journals, manuals, computer screens, and speech are examples of vehicles. Form expresses the various sorts of information which may be derived from the science base. Examples are checklists, user models (eg the Human Information Processor (Card *et al, ibid*)), task models, critiques of dialogues and screens, and guidelines.

This chapter reports work carried out that aimed to improve the dissemination and applicability of research findings from a project concerned with the usability of videotex. An attempt was made to specify the appropriate applications representation for designers of interactive videotex applications. The investigation included an identification of the way these designers use information in making their design decisions. Later expression of videotex research findings (that is 'particularisation' in the terms of Long's framework) in homogeneous terms or in a compatible way would encourage synthesis to occur, that is, inform real world design decisions with research results. Both the teleshopping research (that is, the addition to the science base) and the particularisation of the findings (that is, the transformation to an applications representation) will be addressed.

6.2   The domain

Before going on to the methods employed in the specification of an application representation, it will be useful to outline the relevant domain and the research carried out. The designers' approach will be understood better given a clear view of the domain they will influence.

The domain is the set of tasks addressed by the research and the designers. The domain of this research is known as teleshopping. Teleshopping involves the use of communications technology for selecting, ordering and occasionally also delivery of goods. Many communication systems have been used for teleshopping, such as the telephone and post (mail order). The research focussed on videotex (VT) systems. A review of the various forms of VT may be found in Gilligan and Long (1984). The most typical version is found in the Prestel© system operated by British Telecommunications plc. More complex teleshopping systems, where the public Prestel network is used as a 'gateway' into specialist interactive applications, were addressed by the research, but the emphasis was on the standard VT service. References to VT in the following should be taken to refer to the standard form of Prestel, except where explicitly amplified.

Public VT uses the public switched telephone network (PSTN) to distribute information from central computers to subscribers, and to carry commands and other information from subscribers to the computers. The two-way directionality that the telephone provides is in contrast with the one-way delivery of 'teletext', although the appearance of the screens of both of these systems appear very similar to a user.

In carrying out the tasks of teleshopping, subscribers have to be able to use the basic VT information access functions and also perform the extra tasks associated with remote shopping, such as examining available items, deciding on the suitability and desirability of goods, requesting items and paying for them. Below is a brief description of the tasks facing the teleshopper. The illustrations of VT style screens approximate the appearance of VT screens. The actual appearance of the text and graphics is slightly different – colour is used (impractical to reproduce here), the text font uses a 5 by 7 mosaic, and a 3 by 2 mosiac is used for graphical elements. However, the screen contents accurately represent typical teleshopping frames.

6.2.1   Teleshopping tasks

Users view information in units of frames and pages (the difference is not important here). Pages may be selected by two methods. The first method is by menu selection. A list of choices is offered and users select from the choices by keying the appropriate number (see Figure 6.1).

The other method is by using VT commands. The important symbols in the command language are * and #. *# shows the previous frame. *n# accesses and shows the page n (n is a number up to 10 digits long). In Figure 6.1, the user could have moved from the index to the shopping page by keying in *199#, 199 being the number of the shopping page.

A contrast is made between 'index' pages and 'information' pages on the basis of how much non-routing information is available. Information pages may offer as many or more menu options as so-called index pages.

There is a third class of page which is very important in teleshopping tasks, the Response Frame or RF. The RF allows users to enter and transmit information back to the VT computer, which can then be collected by, or passed onto the Information Provider (or IP). An RF is shown in Figure 6.2.

The RF differs from an information or index page in one major respect.

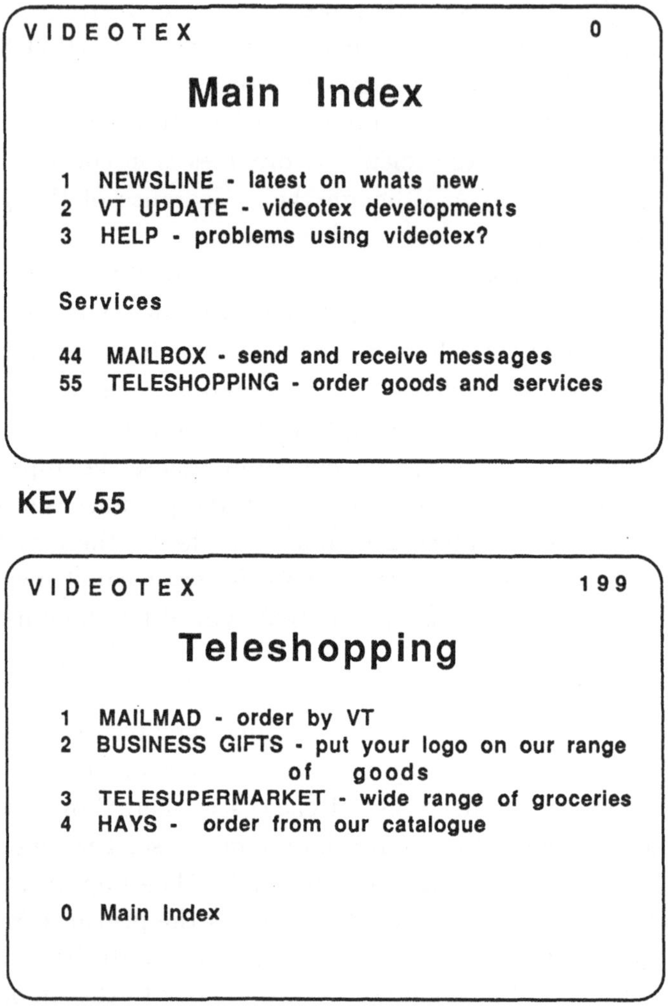

```
┌─────────────────────────────────────────────┐
│  V I D E O T E X                          0   │
│                                               │
│             Main  Index                       │
│                                               │
│                                               │
│     1   NEWSLINE - latest on whats new        │
│     2   VT UPDATE - videotex developments     │
│     3   HELP - problems using videotex?       │
│                                               │
│     Services                                  │
│                                               │
│     44   MAILBOX - send and receive messages  │
│     55   TELESHOPPING - order goods and services │
│                                               │
│                                               │
└─────────────────────────────────────────────┘
```

KEY 55

```
┌─────────────────────────────────────────────┐
│  V I D E O T E X                        199   │
│                                               │
│            Teleshopping                       │
│                                               │
│     1   MAILMAD - order by VT                 │
│     2   BUSINESS GIFTS - put your logo on our range │
│                    of    goods                │
│     3   TELESUPERMARKET - wide range of groceries │
│     4   HAYS -   order from our catalogue     │
│                                               │
│                                               │
│     0   Main Index                            │
│                                               │
└─────────────────────────────────────────────┘
```

Figure 6.1: Menu selection

The RF includes 'fields' which are intended to be filled in by users. In shopping, the fields are used to accept details of the user's orders, the RF being used here as a sort of order form. Figure6.3 shows a completed RF. Users may correct errors as they enter information by keying the command **. This command erases the current field and places a cursor at the beginning of that field. On completion of the last field in an RF a system message is displayed at the bottom of the screen which asks for confirmation to send the RF (as shown on the Figure).

Four types of RF have been identified (Gilligan and Long, 1984; Fenn and Buckley, 1987). These differ in their relationship to information pages and in the completion task required. Figure 6.3

Figure 6.2: A VT style RF

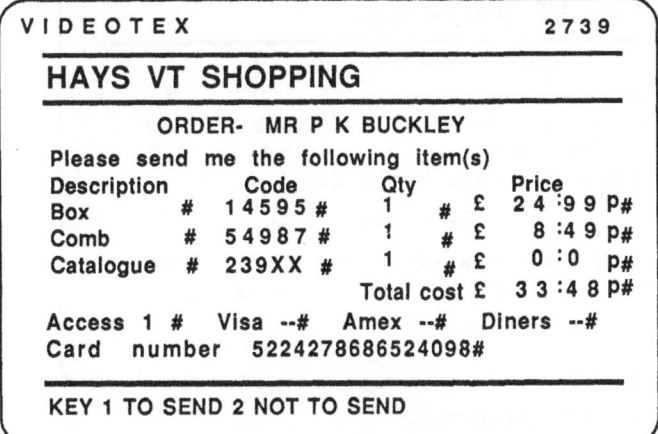

Figure 6.3: The RF completed

shows an example of a so-called generalised RF. It serves as an order form for a possibly large number of different goods, and probably many information pages (given the limitation on text length that pages have). The generalised nature of the RF requires that a number of fields need to be filled in to define the required goods. A variation on the generalised RF is the open RF (Figure 6.4). Only one field is provided, giving a sort of free-format order capability. The other two sorts of RF are more specific to

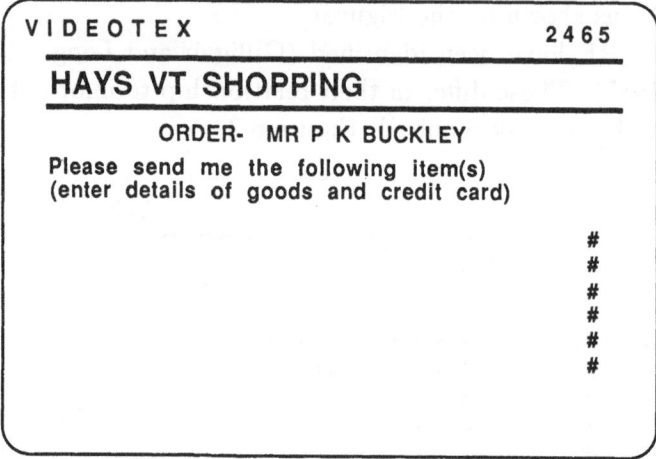

Figure 6.4: VT style open RF

goods, and so are normally linked to less information frames than are the previous two sorts of RF. The menu RF presents the user with a choice of goods (Figure 6.5). The tailored RF is specific to a single item (although

```
VIDEOTEX                              2744

  HAYS VT SHOPPING

          ORDER-  MR P K BUCKLEY
  Please send me:                     Qty
  Nibsons Guide to Fish        £15.99    ---#
  Deep Sea Wonders             £11.45    ---#
  Crustaceans of the Pacific £19.99     ---#

              Total cost      £---:--p#

  Access --#   Visa --#   Amex --#   Diners --#
  Card    number       ----------------#
```

Figure 6.5: VT style menu RF

goods' features such as size and colour may be variable) (see Figure 6.6).

The identification of components of the shopping task in both the shop and VT form was made possible by constructing a model of transaction processing. The model comprises 4 components. The tasks described above, those of filling in RFs are part of the task component of display – the shopper has to specify their requirements (the goods ordered) and their own payment to the shop or other seller. Other task components are acquisition – obtaining information about goods and their costs, and evaluation – assessment of goods (through their representation) against shoppers' criteria for those goods. The fourth component, exchange, represents the transfer of ownership from one transactor to another.

Acquisition tasks are the tasks required to effect changes of viewed page, and associated navigation through the data. One such task would be to select the appropriate menu option for information about the required goods. This task is easy or difficult depending on the ease of categorising one's target goods in the options provided. For instance, given the options in Figure 6.7, it is a simple matter to choose 3 BOOKS if the target is a book. On the other hand, if the target is a television, it is more difficult to identify the appropriate category. Is it 4 ELECTRICAL GOODS, or 5 HOUSEHOLD GOODS?

Evaluation is the final task component to be described here. Text descriptions of goods tend to be brief and graphic descriptions (if any) tend to be crude, for example as in Figure 6.8. Whereas in a shop one can normally request further information on many details of the goods and the transaction (and in most cases have a richer representation of the goods anyway), on VT the representations are simple, and only pre-specified information is available. If one's query is not addressed in the information pages, one

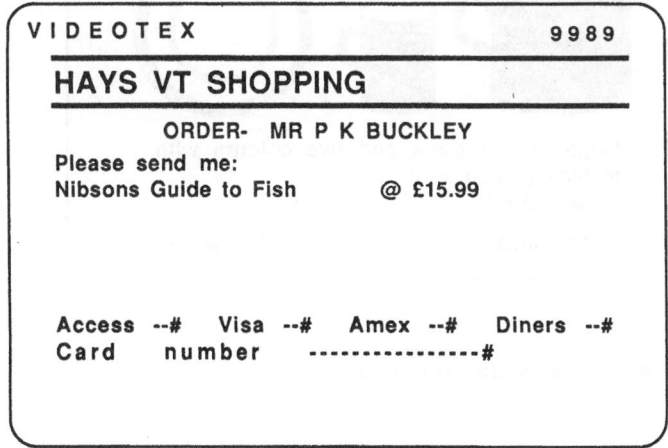

Figure 6.6: VT style tailored RF

```
VIDEOTEX                               9900

HAYS VT SHOPPING index

CHOOSE FROM THIS LIST

1 CLOTHING - men's and women's
2 FOOD - groceries and gifts
3 BOOKS - novels and non-fiction
4 ELECTRICAL GOODS - fridges etc
5 HOUSEHOLD GOODS - brooms, storage systems
6 FURNITURE - for the bedroom or lounge


9 HAYS shopping          0 Main index
```

Figure 6.7: Teleshopping style goods menu

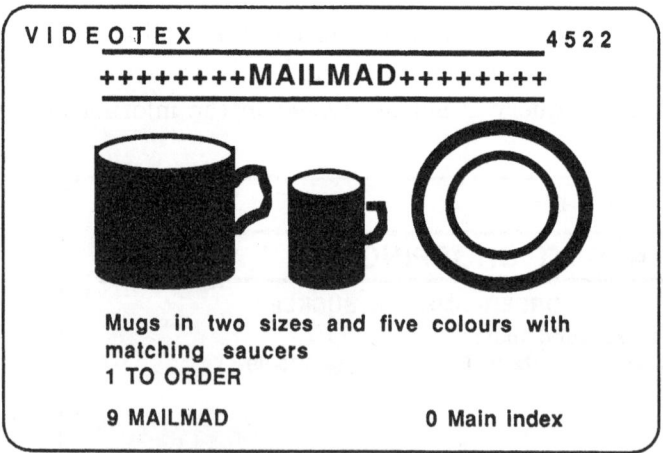

```
VIDEOTEX                               4522
++++++++MAILMAD++++++++




           Mugs in two sizes and five colours with
           matching saucers
           1 TO ORDER
           9 MAILMAD              0 Main index
```

Figure 6.8: VT style illustration of goods

has to use some other information source.

The designers of the VT dialogues make decisions that affect the characteristics of the teleshopping task described above, such as the type of RF, the database structure and the names of the menu options. The next section outlines the teleshopping research carried out which provided information about the behavioural consequences of various dialogue designs. Later in this chapter, the work of the designers will be examined in more detail as part of the attempt to identify the appropriate expression of the research findings.

6.3   The research

The research had three main phases. The first phase defined the general domain and the particular technology to investigate. As can be gathered from above, the domain chosen was transaction processing using the Prestel version of videotex technology.

The second phase was the identification of problems teleshoppers might face. This was done by, first, observation of inexperienced VT users attempting to shop using a real public teleshopping service, and second, an analysis of the difficulties they experienced, and the errors they made. The analysis provisionally identified system features and inappropriate user knowledge that was associated with the difficulties and errors. The third phase of the project sought to confirm the features as sources of user error and difficulty. A series of controlled experiments was carried out in which system and knowledge variables were manipulated.

The research findings are, then, of two sorts, corresponding to work done in the second and third phases of the project. The second phase supplied a list of system features that appeared to contribute to users' problems. This list was expressed as a set of system variables and associated with a set of knowledge variables in a model derived from the Block Interaction Model (BIM) (Morton, Barnard, Hammond and Long, 1979). The model expresses difficulties as a mismatch of the ideal user knowledge and the actual recruited user knowledge. Technology itself defines the ideal (error-free) operations needed to make it work. Therefore, the ideal knowledge was expressed in terms of system features. The information from this phase that might be useful for designers is the corpus of errors, and the potential source of errors, the interaction of the system and knowledge variables. However, these results describe only potential sources. There was at this stage no rating of the seriousness of the error, nor any validation of the Interaction model.

In the third phase a series of experiments examined some of the variables identified earlier. With the results, the effects of system features on user performance could be confirmed and quantified. However, only a small subset of the potential combination of system and knowledge variables could be examined within the resources of the project. The third phase lacks range, but the results have greater validity than the hypothesised effects described in the second phase.

6.3.1   Summary of results

The results are summarised below. In the second phase, the analysis indicated that the following system variables had an effect on user difficulty and error rate. The variables are defined in Buckley and Long (1988).

> Jargon
> Command Language
> Screen (graphical characteristics such as colour combinations)
> Organisation of System
> Instructions
> Range of Goods
> Extent of Description of Goods
> Modality of Description of Goods
> Organisation of Goods
> Input Device
> Input Demands
> Response Frame Type
> Delivery Delay
> Availability of Goods

We needed also to identify knowledge variables, since our basic model of the sources of difficulty required the identification of two factors – the ideal knowledge, and the actual mismatching knowledge. The system variable represents the ideal knowledge. The user knowledge that appeared to be inappropriately recruited in VT shopping tasks was expressed as knowledge sources. The list of knowledge sources below was constructed:

> Knowledge of Brands
> General Knowledge of Transaction Domain
> Workbase Knowledge of Transaction Domain
> Knowledge of Natural Language
> Knowledge of Computer Systems
> Knowledge of Off line Systems
> General Knowledge of Transactions

Definitions of the variables are given in Buckley and Long (1988).

Particular knowledge variables and system variables were associated as pairs with particular evidence of difficulty in the record of the initial observational study. In this way, the source of the difficulty (or error) was described as the mismatch of the knowledge required for the system variable, and the actual knowledge the naïve subjects used – the knowledge variable. The combination of the two sorts of variable constituted a 'hypothetical sources of difficulty' model. The third phase sought to confirm aspects of this model.

The experiments tested the following:

1. The representation of goods, and the knowledge of goods for the task component of evaluation (Buckley and Long, 1986)
2. The RF type and input demands for the task component of display (Fenn, 1985)
3. The RF type and input demands, and the knowledge of transactions, for the task component of display (Fenn and Buckley, 1988)

A literature review was used to collect evidence for the usability of VT for the acquisition task component. This component is regarded as having few differences with the use of VT for information retrieval in general, and thus research carried out elsewhere would be appropriate for teleshopping (eg the modelling of menu choice by MacGregor, Lee and Lam, 1986).

The research findings related behavioural data and an interpretive model (the BIM) to features of the teleshopping task. The lists of variables indicate potentially critical usability features, and the experimental results provide validation and a measure of the strength of their effects. These constitute candidate additions to the science base in the terms of the HCI framework, and potentially useful information for the purposes of improving VT dialogues.

6.4   The teleshopping dialogue design

In order that VT dialogue design can incorporate this information, our findings need to be communicated to the dialogue designers. In terms of the HCI framework, we want to aid synthesis. The design activity offers a potential route for the synthesis of research findings with real world technology. As synthesis is assumed to be a designer's responsibility in this domain, our favoured strategy is to provide designers with an appropriate applications representation. The form and vehicle of this representation should be appropriate in that they are homogeneous or compatible with the designers' current use of design advice. An analysis of the design activity should provide information about the designers' work, including 'synthesis' activity, that is their synthesising of information from the science base with decision making in VT dialogue design.

The next part of this chapter is about an empirical examination of the activities associated with teleshopping dialogue design and construction. If we can understand how teleshopping dialogues are constructed, and by whom, we are better able to specify the sort of information that should be included in an application representation (the form), and the way the information should be presented to designers (the vehicle). Later, it is important to particularise the research findings described in Section 6.3 into the specified representation, although this step has not yet been carried out.

6.4.1   The designers

The results of the research and the associated design suggestions were disseminated to the research community through the normal channels such as academic journals and conference proceedings (see reference list). However, it was impossible to know if the information affected, or even reached the people responsible for the design of the teleshopping dialogues. This section is about the identification of the relevant group of designers among the numerous sources of variation in a teleshopping system. A later section will discuss the use of information by the identified designers.

There are many sources of influence on the design of a teleshopping system. The videotex standard specifies the protocols that determine the coding of the pages, including how attributes such as colour and size are represented. The actual colour, character size and shape displayed are independent of the standard, and rely to a large extent on the decisions of the designers of the display equipment, although in practice most videotex terminals behave more or less the same at the display level. The provision of information and interactive services on public videotex systems is the responsibility of a number of groups. The *operator*, in the case of Prestel, British Telecommunications plc (BT), provides the storage and communication infrastructure as well as some high-level structuring of information. Typically, the 'root' indexes are designed and maintained by the operator.

The great majority of the videotex data is provided by more or less independent *Information Providers* (IPs). The IPs lease storage space from the operator and then have a responsibility to provide information services. To a large extent the IPs are autonomous. The operator's leasing policy determines the costs to the IPs of providing videotex services. During the period of the research, the policy of BT was for IPs to lease a minimum of 100 frames. This proved to be more than most IPs needed themselves, and so they in turn sub-leased space to *sub-IPs*, whilst keeping the frame and data structure design and maintenance responsibility. The *full-IP*, then, appeared to be responsible for most decisions that affected the design of information and interactive services on public videotex. Also, the full-IP has responsibility for the type of dialogue feature open to design changes (ie they design the information provision and frame routing, rather than define videotex Standards).

The types of system feature that might be expected to affect performance had, of course, been identified in the earlier research. We had listed 14 system variables that appeared to affect usability. Which of the design groups described above had responsibility for these features? It appeared that the full-IPs were the main source of usability-related variation in videotex information and interactive services. Having identified the target group of designers, the next stage was to find out how they approached their task, and specifically how they designed the usability-affecting features. This information we thought would be useful in identifying the appropriate vehicle and form for the applications representation.

6.4.2   The design tasks

Information about the ways in which the full-IPs created information and interactive services dialogues was acquired by interviewing a small sample of experienced videotex designers. All the informants had designed interactive dialogues that exploited one or more response frames. Only a small sample was used (five designers at four different companies) because of limited time available, and the low priority of this work element in the project. A larger sample would have been more useful, and future projects that present themselves as applied might with advantage plan more resources for investigation of just how the results will impinge on design effort. The rest of this section is a description of the task of the videotex dialogue designer synthesised from the interviews with the sample.

As most of the design work of the full-IP is on behalf of the sub-IP, one early part of the task is to identify what the client wants presented on the system. Often, the sub-IP has a view of the capabilities of videotex, and initially presents paper versions of the information they want the full-IP to mount. Sometimes, the view is uninformed, and the type of material offered is, according to the VT designers, unsuitable. This is illustrated by the experience of one designer who was presented with a printed colour brochure outlining the services of a company, and then asked to 'put it on the system', as if a direct reproduction of the pages of the brochure was required. The designers felt that information has to be designed specifically for videotex presentation, and the initial part of their task is in convincing clients to agree with them. Sometimes their efforts fail, and the client insists on presenting what the designers see as inappropriate material, such as long detailed text passages, and complex illustrations. Also included in the preliminary discussion is the space requirement of the client. The designer and the client will together decide how many pages and response frames the information provision will use, in the light of financial constraints from the client, and technical constraints from the system (eg the text capacity of frames).

After the preliminary discussions about the design of the dialogue, the designer will create the dialogue. There are a number of ways in which this is done. Most of the work is the production of small and simple dialogues, such as the design and routing of a set of 10 VT pages, comprising an initial index page, 8 different information pages, each perhaps describing a different product, and a single response frame. In such cases, the designers will often enter the frames and the routing information on-line to the operator's machines. A similar alternative is to input to a local system, and up-load the pages at a convenient time. For more complex work, which is less common, the designer may plan out the pages and their relationships on paper. One example of the more complex presentation was the provision of a house price information service for a building society. The designer was given a brochure which tabled information about how house prices in different regions changed over time. The design task was to provide a videotex database that provided this information. The designer used a large sheet of paper to work out the links in a database which allowed the user to access the information in varied ways (for example by region, then by year and then by house type), and then allow easy comparison with other regions. Once the paper plan was finished, this designer then entered the pages and routing information on-line. The graphic design of the pages, rather than the routing design, is almost invariably carried out on-line. The few exceptions occur when a client requires complex illustrations. Videotex artwork is generally provided by specialists, not the dialogue designer.

In most cases, the design task is completed when the dialogue is linked into the public database. In some instances, however, the designers tested out the dialogue, either on themselves, or sometimes with a group of subjects. Feedback from public users is available, and occasionally this route is used to bring designers' attention to errors in the dialogue, such as non-existent menu options, or loops. Feedback is not formalised. Indeed, while being interviewed, one designer demonstrated a new dialogue he had mounted and discovered that a menu option led to an incorrect page. In this case, and generally, designers correct errors on-line. This is possible because of the simplicity of most of the designs. The operator also monitors the system, and can request alterations to the design of pages and database structure.

6.5   Analysis of design tasks

The above has described the identification of the designers and provided a summary of the activities of these designers. This section attempts an analysis of these activities. We are particularly interested in how the designers apply human factors information to their design problems (if at all), since our intention is to add to this application. The analysis will be in terms of designers' 'resources' and associated synthesising 'transformation' into the dialogue features.

6.5.1   Resources

What sorts of information do designers use currently? The experienced sample relied to a large extent on what could be termed internal resources, such as knowledge of the client, and memory of instances of their own or others' VT design. These resources are in contrast with external resources such as journals, manuals and other reference material, and existing examples of VT dialogues.

First, the use of external resources. In three of the four establishments visited, the interviewed designers had contributed sections to a 'house style' manual. However, they did not consult this manual when developing and building teleshopping dialogues. A printed IP manual is provided by the operator, but this was never used or even read by the designers. Indeed, one informant expressed surprise at the interview when told that the IP manual included a section of advice on dialogue design. Of those designers who had at one time examined the IP manual, the consensus appeared to be that the information it contained was too general to be of use. A handbook (the 'Code of Practice') published by an industry authority (the Videotex Industry Association) was used occasionally for the specific advice it provided about current codes of practice concerning, for instance, charging users for page accesses, and the importance of making clear a contrast between 'information' and 'advertising'. The handbook includes far more material on the regulations governing 'advertising', 'consumer credit', 'claims for slimming products' and so on, than material on dialogue design, such as the guidelines for information completeness and routing. There is no information on frame design in the handbook. Clients' printed material is used as a basis for the dialogue to be constructed (eg the house prices brochure). The designers did not use psychological or ergonomic reports, and one of them did not realise that research in this area was carried out and published.

Designers seem interested in the work of other designers. They occasionally reviewed the state of the public databases, and assessed the dialogues, criticising what they thought were poor designs, and assimilating and subsequently exploiting what they thought were good solutions. They were generally able to identify the author of dialogues they tried out on the live system. This was possible because each designer had a personal style and the VT designer community was then quite small. The system is used as a resource in a number of other ways. A section of the VT database provides advice for dialogue design. The system also offers feedback; the operator will sanction designers if they break the regulations for VT material (these cover honesty and decency issues); also IPs can utilise a frame access counter to provide feedback about the number of times a particular frame is accessed by a user. An abnormally low number of accesses may be due to poor routing and frame design. Feedback is also available from users, who may contact IPs to complain about poor routing and layout. Very occasionally, for complex systems, designers set up user trials and the results of these provide useful information about the acceptability of the designs.

There are a number of established conferences for the VT industry, and designers do attend these, although the material presented is regarded by them as not very useful. BT occasionally runs seminars, and these are attended by the designers. None of the designers attended ergonomics or HCI conferences.

It is more difficult to identify and characterise internal resources, although it is clear that these are valuable for the designers. For the more frequent simple design tasks, designers tended to identify and use solutions they had used previously, such as the 10 page set described above. They did not consult the previously designed dialogues, or any materials apart from clients' documents, and so appeared to have had the solution as a knowledge resource. Design knowledge and skill seemed to be a source of pride to them, and they frequently expressed how important experience was for them in its development.

Their rationale for the solutions, elicited in the interviews, occasionally referred to the effects of dialogue features on users – for instance, 'they don't like finding dead ends'. This could be taken as weak evidence that the designers have and exploit a rudimentary user model. As well as referring to the users' affective responses, the designers also addressed performance time. It was clear to them that a user wants to find the relevant information as quickly as possible. However, there was no quantitative analysis of dialogues, for example in the manner proposed by Card, Moran and Newell, 1983; there was no evidence of calculated evaluation of usability. Occasionally, for the more complex design tasks, the designers reported that they attempt to imagine what the user is trying to do. In terms of resources, this requires the designers to utilise some sort of task model.

In summary, the designers rely largely on knowledge of standard solutions. They have a vague view of the behavioural consequences for users of these solutions. A number of external resources, such as manuals, record the favoured approach of the experienced designers. However, the manuals are not normally used as a resource during design tasks. The next section describes how the designers utilise their various resources in the solution of a design problem.

6.5.2   Transformation (Synthesis)

The designers expressed concern about the ease of use of the dialogues, and had some clear views of how some system features under their control would affect users. As an example of this, they all agreed that text should be presented as complete units within pages, rather than split between a number of pages. But they did not report any use of traditional forms of human factors information, which are expressions of the science base, and are normally represented in research papers and design handbooks. Because of this it was difficult to identify synthesising transformations of human factors information to the dialogues. However, it was possible to examine the ways in which designers influenced the parts of the system that the research had identified previously as being potential sources of user difficulty and error (the system variables). Some of the parts that affected usability were either fixed, or otherwise not under the control of the full-IP. These were essentially the Command Language, and those features covered by the 'goods' variables – Range and Availability of Goods, and Delivery Delay.

Specific features of the system were on occasion determined in isolation, but the standard solutions include specifications of a range of features. The only external resource normally used here is information from the client that forms the basis of the material to be expressed as a VT database. Internal resources include the framework of a complete simple solution (in terms of VT features), into which specific client information will be inserted, and knowledge of the house style. In more complex design tasks, more decisions will have to be made. On occasions, the designer's decision is aided and justified with reference to the imagined effects on users. As mentioned above, this could be regarded as a synthesis of informal task and user models with characteristics of the VT dialogue.

In the example of the house price information service, the designer drew the structure of the data on paper before constructing the database. The diagram also included information about the menu options that would appear to the user. In the interview, the designer of this database referred explicitly to his attempt to 'imagine what the user is trying to do'. This task model is not based on empirical work – task analysis, for instance, but only on the designer's imagination. External resources, then, are used rarely during the construction of dialogues. Rather, internal resources such as user and task models and knowledge of previous design solutions are used to design and construct the routing and the appearance of the frames.

Clearly, the design behaviour identified in this observational study is different from the design handbook scenario described in the Introduction. The design handbook may have a function for the VT designers, but not as a tool for the calculation of usability of system options. The only quantitative data the VT designers use are frame accesses, and the design and construction of dialogues is frequently carried out in a single session using a pre-existing internalised framework. The analysis indicates that in the design and construction of particular VT dialogues there is no use of external resources that could be identified as an applications representation. Instead, the designers rely on internal resources such as solution frameworks and rather vague models of users and tasks. For our purposes in encouraging synthesis of research results with the VT design, it seems more important to choose an application representation that will affect the development and content of these internal resources, rather than add to the already quite extensive set of infrequently used IP manuals. As part of the specification of an appropriate applications representation, the next section describes some possible vehicles and forms, and assesses whether they could be regarded as compatible with the VT designers' use of resources.

6.6   Applications representation
6.6.1   Vehicle

Clearly the research published in conventional human factors vehicles would be unlikely to reach the attention of the VT designers. Other paper vehicles are also poor candidates. The Prestel IP manual appeared originally as a likely vehicle for the applications representation, but the evidence from the designers showed that they did not refer to or use this material. (However, the manual does serve as a sort of archive of current VT design practice, and the research information could be expressed here as additional entries.) It appears likely that the sample of designers would not be informed if the IP manual carried the only applications representation. House style manuals are also used as an archive of established design practice, rather than as a 'live' tool, and so the same criticisms apply. The Videotex Industry Association handbook is more frequently used, but as a source of information about advertising standards and other publishing legislation. The dialogue design guidelines in this handbook again are to these designers a record, rather than a design tool.

One of the more influential resources, perhaps because of its immediacy and frequency of use, was the live Prestel system. From this resource, designers picked up ideas from other designers through the use of their databases. As well as being popular, the Prestel system has the additional advantage of being able to show the results of choosing among different design alternatives.

VT dialogue designers will pick up information presented at industry events, for instance conferences, and IP seminars. These provide various situations in which designers will be exposed to information about VT use, and provide various reasons for elaborating or otherwise modifying their task and user models.

6.6.2   Form

The information in the science base, such as research results, influences both the possible forms and vehicles of the applications representation. For instance, the results outlined here could not be used directly to generate a predictive user model, in the manner of the GOMS models (Card, Moran and Newell, 1983). This section discusses possible forms for the applications representation, given the information in the science base provided by the teleshopping research, and the preferable forms, given the analysis of the designers' tasks.

The most obvious possibilities for the form of expression of the research findings follow the original research intentions. The system and knowledge variables identified in the observational study provide a list of features that were informally associated with user difficulty and error. The two types of variable may be separated to provide two levels of user modelling. The system variables alone indicate the system features which are usability-sensitive in an undefined population of users. When the knowledge variables are added, user characteristics are defined, and so a user population may be structured. The model of the sources of difficulty, based on the interactions of knowledge and system variables, could provide an indication to the designer of the circumstances (the potential user group) under which particular system features become critical. The results of the later experiments confirm some of these interactions, and also provide some further information. As an example of confirmation, the evaluation experiment (Buckley and Long, 1985b) confirmed the value of Knowledge of Brands. As an example of extension, the RF experiment showed that the performance and user acceptability of the different types of RF both varied given the type of shopping task being carried out. For single items, the tailored RF was clearly superior, but for orders of several items the generalised RF was better (Fenn and Buckley, 1988).

Although it is of course possible to express this model as a series of checklists of critical features, each corresponding to an identifiable category of users, it appears from the analysis that the designers do not use such external resources. The form should preferably be designed to be used and assimilated before the dialogue design and construction, in effect to become part of the designer's user model. The identification of the particular system features in the application representation is important since the designers claim to reject over-generalised information, as is found in some of the design manuals.

One quite popular information resource for the designers was the live Prestel system. Designers examined others' interactive and information databases to criticise and pick up ideas. This vehicle could be used to represent in a text form the description of the user model and the list of critical features. This would combine the vehicle of VT with the form of a research paper. However, a better exploitation of this popular vehicle would be to mount a series of dialogues showing situations in which the sources of difficulty model would point to potential and confirmed difficulties. The system could show the actual system feature in the context of a simulated teleshopping dialogue. This has an advantage of providing concrete examples of the critical features, rather than descriptions of the features, and is as a result less generalised. Expression of the research results in this vehicle and form is expected to provide an influence on VT designers. The current expression of the findings as research papers has little likelihood of impact on real VT design, as it is expressed in a vehicle which does not constitute a resource for the relevant designers.

6.7   Summary and conclusions

The work of VT designers was analysed in an attempt to specify the vehicle and form of a representation of results of research that had addressed the designers' domain. The main reason for the analysis was the belief that the transfer of research information from its normal expression in the 'science base' to an influential status in design work, requires the information to be expressed in a way suitable for designers. In order to identify the most suitable expressions, it was felt that it was important to find out how designers currently carried out dialogue design, particularly what resources they used and what transformations they carried out to define features that affected usability. The designers themselves were assumed to be the best source of information about their design task behaviour, and so a sample of them were interviewed. The interviews gave an empirical base to an analysis of the design resources normally used by VT dialogue designers. A final step is to adapt the sorts of research information that had been developed to the vehicles and forms of expression that had been found to be used by and useful to designers. Whereas the range of potential vehicles and forms has been identified and described above, the research findings have only been expressed as standard research reports in conference proceedings and academic journals.

The observational study of the work of the VT dialogue designers and the analysis of this work in terms of resources and transformations suggests that the designers rely on internal information resources that are characterised by specific system features, such as a particular set of frames with particular relationships. They do not use external resources such as design manuals, and resort to rough but generalisable user and task models only if the design brief cannot be solved with a particular standard solution. An application representation that appears to be compatible with this approach to a design solution would combine the vehicle of a VT database with the form of dialogue scenarios, with indications of the user-sensitive system features on which the scenarios are contrasted.

6.7.1   Value of approach

In both the compatibility approach taken here and the homogeneity philosophy of Card *et al* (1983) the designer is the central agent of research application, and it is necessary to adapt information to suit their current usage patterns. There is an inherent weakness in that the approaches are purely reactive to a current situation. In some circumstances, designers do not use information from the science base, or deliberately avoid consideration of some areas. This may be because of weaknesses in the coverage of the science base, scepticism arising from poor results of the use of earlier applications representations, or perhaps information masquerading as such.

The evidence from the interviews suggested that the VT designers were in fact concerned with the effects of their solutions upon the users, and would welcome information that helped them in this respect. This attitude might have been encouraged by some characteristics of the VT system: the designers had to cope with the consequences of a public system, that is a potentially highly variable user population, with a large proportion of casual users. This is in contrast to most applications, which are destined to constitute part of the technological environment of a particular workforce. In addition, the designers regarded the VT domain including teleshopping as a specialised form of publishing rather than of computer interface design, and were consequently concerned to attract and maintain the interest of the user. Operational difficulties for the user were seen as detrimental to the achievement of this aim.

Given these designers' concerns, the compatibility approach is probably acceptable. In other circumstances, work may be required to find ways of encouraging the designers to consider the user to a greater extent. As an example, the Programmable User Model (PUM) project (Young, 1985) is exploring a way of directly affecting the design by putting the designer in what effectively constitutes a negative feedback loop. Deviations of the design from that matching the abilities of the user model are explicitly discouraged or rejected.

The characteristics of the design tasks were identified on the basis of an empirical study. Similar work will be required for other projects where the application of research results is of concern. The amount of time and effort for this component in the VT work could have been increased with advantage. The main influences on the interface, those resulting from decisions of the full-IP, were captured, but other variables, especially the goods variables, were not covered. Also, although it was possible to specify the vehicle and form for the applications representation, there was no time to express the research results in this way.

6.7.2   *Lessons learnt*

At the conclusion of the study it was clear that it is of little use in this domain simply to publish in the standard science base record (the scientific journals), and trust that the implications for design will be picked up and synthesised in the design function. The contact with the designers provided some insight into the sorts of issues they find important in resolving design decisions, and their use of various resources in this work. The designers' behaviour could be used as cues for initiating research into specific design questions, but care should be taken that the requirements of the science base are also considered, otherwise research could be in danger of providing over-specific information isolated from the generalising capabilities of theory.

From an empirical study of the type described here, to the specification of an appropriate applications representation, requires extensive work. Here it was found that VT dialogue designers relied to a far greater extent on internal resources than on external (and therefore easy to identify) resources. Other attempts might with advantage include a validation phase to check the assumptions about the nature of the internal resources. Design in other domains may well follow the scenario offered by Card *et al* (1983), and rely to a greater extent on the external resources.

The teleshopping research findings have not been expressed in any form except as research reports. It is not yet possible to report on the success of the approach taken, either as an aid to the dissemination of results to the design activity, or as an influence on the usability of shopping dialogues.

However, it has been possible to show the weaknesses of current vehicles and forms of representations of research findings, and to provide empirically based suggestions for improvements.

Acknowledgements

This work was carried out while the author was at the Ergonomics Unit, Department of Psychology, University College London. The teleshopping project described here was funded jointly by British Telecommunications plc and the SERC under grant GR/C/23032.

References

Buckley, P.K. (1985); Realising the potential of viewdata, in N. Bevan and D. Murray (eds), *Man-machine Integration*, Pergamon State of the Art Report, 13/1, Pergamon Infotech, Maidenhead

Buckley, P.K. and Long, J.B. (1985a); Identifying usability variables for teleshopping, in D. Oborne (ed) *Contemporary Ergonomics 1985*, Proceedings of the 1985 Conference of the Ergonomics Society, Taylor and Francis, Basingstoke

Buckley, P.K. and Long, J.B. (1985b); Effects of system and knowledge variables on a task component of teleshopping, in P. Johnson and S. Cook (eds) *People and Computers: Designing the Interface*, Cambridge University Press, Cambridge

Buckley, P.K. and Long, J.B. (1986); Recommendations for optimising the design of teleshopping dialogues, in D. Oborne (ed) *Contemporary Ergonomics 1986*, Proceedings of the 1986 Conference of the Ergonomics Society, Taylor and Francis, Basingstoke

Buckley, P.K. and Long, J.B. (1988); Using videotex for shopping – A qualitative analysis, *Behaviour and Information Technology*, paper accepted for publication

Card, S., Moran, T. and Newell, A. (1983); *The Psychology of Human Computer Interaction*, Academic Press

Fenn, P.S. (1985); *A comparison of the response frames used in teleshopping*, unpublished MSc dissertation, University of London

Fenn, P.S. and Buckley, P.K. (1987); Using videotex to order goods from home, in E. Megaw (ed) *Contemporary Ergonomics 1987*, Proceedings of the 1987 Conference of the Ergonomics Society, Taylor and Francis, Basingstoke

Fenn, P.S. and Buckley, P.K. (1988); Just fill in the details: a comparison of the response frames used in teleshopping, *Behaviour and Information Technology*, paper accepted for publication

Galer, M. and Russell, A.J. (1987); The presentation of human factors to designers of IT products, in Bullinger, H.J. and Shackel, B. (eds), *Interact 87*, North-Holland

Gilligan, P. and Long, J.B. (1984); Videotext technology: an overview with special reference to transaction processing as an interactive service or Everything you never wanted to know about text on TV and were terrified that someone would tell you, *Behaviour and Information Technology*, 3, No.1

Long, J.B. and Buckley, P.K. (1984); Transaction processing using videotex or Shopping on PRESTEL, in Shackel (ed), *Interact '84*, Proceedings of the first IFIP conference on human-computer interaction, North-Holland

Long, J.B. and Buckley, P.K. (1987); Cognitive optimisation of videotex dialogues – a formal-empirical approach, in M. Frese, E. Ulich, W. Dzida (eds) *Psychological Issues of Human Computer Interaction in the Work Place*, Elsevier, North-Holland

MacGregor, J., Lee, E. and Lam, N. (1986); Optimising the structure of menu indexes: a decision model of menu search, in D. Oborne (ed) *Contemporary Ergonomics 1986*, Proceedings of the 1986 Conference of the Ergonomics Society, Taylor and Francis, Basingstoke

Morton, J., Barnard, P., Hammond, N. and Long, J.B. (1979); Interacting with the computer: a framework, in E.J. Boutmy and A. Danthine (eds) *Teleinformatics '79*, Elsevier, North-Holland

Young, R. (1985); position paper on 'User models as design tools for software engineers', Alvey workshop on MMI/SE, September 1985

7

Task analysis for knowledge descriptions: theory and application in training

Dan Diaper,

Peter Johnson

Summary

The development and application of the Task Analysis for Knowledge Descriptions (TAKD) methodology to syllabus design is described. The stages of TAKD are exposed and related to Long's (1986) framework for cognitive ergonomics. The supporting psychological theory of human knowledge underlying the application is described. The more general application of TAKD is discussed with respect to the design of computer systems.

7.1   Introduction

The main body of this chapter describes the application of Task Analysis for Knowledge Descriptions (TAKD) to the design of a training syllabus in Information Technology (IT). This application is used to describe the method in sufficient detail that its potential will be clear to others interested in applying TAKD. The various stages of TAKD are related to the acquisition and application representations of Long's framework for cognitive ergonomics (1986, 1987). The use of this framework exposes the various products of TAKD at different stages of its application. The more general application of TAKD to Human-Computer Interaction (HCI) research is discussed in the final section.

7.1.1   Background to the research

The major part of the research described here was carried out between 1982 and 1985 at the Ergonomics Unit, University College London with funding from the Manpower Services Commission (MSC). The project's remit was to develop a syllabus for training in IT for school leavers on the Youth Training Scheme (YTS). It was decided early in the project that the syllabus should represent the knowledge necessary to perform IT tasks, rather than tasks themselves, to allow for the diversity of application that would be required of the implemented syllabus.

In early 1985 the MSC supported almost 200 Information Technology Centres (ITeCs) as part of the YTS. Each ITeC had between 30 and 90 trainees with a trainee-staff ratio of approximately five to one. Trainees were typically 16 year olds who had left school with few or no academic qualifications. At this time, ITeCs offered a one year training course in three broad areas of IT: microelectronics; computing; and automated office applications. Training in computing involved the development of basic programming skills (often, but not always, in the language BASIC); in microelectronics it included circuit design, construction and testing and also machine code programming and some aspects of robotics; automated office applications covered word processing, accounts packages, data-base applications and non-computerised applications such as the use of telephones, photocopiers and general office procedures. Keyboard skills were often taught using simple Computer Assisted Learning packages. At the time of the research project, there was no agreed syllabus for ITeCs and each developed its own training courses, though some fragmentary guidance was provided by the MSC and from well established ITeCs.

The MSC had decided that ITeCs would concentrate on hands-on teaching methods, as opposed to classroom, lecture-style teaching, as many trainees had not done well within the school environment and it was believed that another year of training similar to that provided by schools was not desirable. The ITeCs explicitly tried to familiarise trainees with a work environment by organising themselves as if each ITeC was a small company. Across the ITeCs, there was considerable diversity in the style, equipment (both hardware and software), personnel and other resources enjoyed. The MSC viewed such diversity as desirable in that ITeCs could thus be established to meet the needs of their local community. Such diversity, however, naturally produced considerable differences in course content and in the tasks used to provide hands-on training.

The syllabus for a one year ITeC course developed from this project was designed to provide guidance to staff in ITeCs and to allow the MSC some central control of course content. It was supposed that a syllabus consisting of specific training tasks would be inappropriate as such task specifications tend to be either too specific (i.e. at too low a level of description) and thus cannot be applied uniformly across the variety of equipment found in ITeCs, or too general (i.e. at too high a level of description) such that there is insufficient detail to ensure the required consistency of course content across ITeCs. To take this latter case, it is insufficient, for example, to specify only that trainees are able to use a word processor as this can mean anything from being able to simply print out a pre-prepared letter on a home microcomputer to producing and editing a multi-media document on a sophisticated document processor. The solution to the problem of diversity was for the syllabus to specify the knowledge required to perform IT tasks, independently of either particular equipment or tasks. The syllabus was to be a listing of the knowledge requirements that would, in combination, be able to describe any IT task.

From interviewing managers, trainers, syllabus designers and examiners, both within ITeCs and in other organisations, it was clear that the content of IT courses relied on the experience, knowledge and skills of either a single person or a small committee. This highly subjective basis for course design was considered undesirable, particularly as the field of IT is undergoing rapid change with, for example, the widespread introduction of quite sophisticated, inexpensive microcomputers. Thus, an individual's personal expertise is likely to become quickly out of date. In addition to the criterion of coping with diversity, a second criterion was established that the syllabus would have a more empirical, less subjective, basis for its content. It was considered highly desirable that the syllabus' content should match IT practice in industry and commerce. Given the changing nature of IT, a third criterion was that the syllabus would be able to accommodate future changes in IT relatively easily and that there should be a mechanism by which it was possible to check that the syllabus had not become out of date.

7.1.2   *Long's framework for cognitive ergonomics*
The original purpose of Long's framework (1986 and also described in this book) was to compare different approaches to cognitive ergonomics, particularly in the field of HCI, though the framework has been extended to characterise and contrast the paradigms of engineering, educational and clinical psychology (Long, 1987). Two "worlds" are hypothesised, the real world, which contains the objects of everyday experience, and the representational world. This latter world can be further divided into two types: (i) science representations; and (ii) intermediary representations. Intermediary representations are used to associate the real world with scientific knowledge and are the principal focus of this chapter. There are two basic types of intermediary representation: (i) acquisition representations; and (ii) application representations. For present purposes, the following slightly modified version of the framework will be used. An acquisition representation provides a description of the real world that is consistent with scientific knowledge and is related to the two by functions that are bi-directional. The "analyse" function provides the real world input to the acquisition representation and is bi-directional in the sense that analysis of

the real world is usually iterative (i.e. the description of the real world is refined over a number of cycles). Simultaneously with this "analyse" function there operates a "particularise" function from the science representation and a "generalise" function to the science representation. The application representations contain the prototype products that are created from the final acquisition representation and are consistent with the science representation via a "predict" function. Each acquisition representation is mapped back to the real world via the "synthesise" function. Again such functions, particularly the synthesise one, will usually be used iteratively.

The TAKD method, applied to the syllabus creation research described here, will use this modified version of Long's framework to expose the number of acquisition representations that TAKD generates and how these are related both to the real world and to scientific knowledge (in this case a psychological model of knowledge representation). Similarly the framework will clarify the development of the application representations and how these are mapped back on to the real world and evaluated.

7.1.3   Overview of the chapter

The chapter is divided into ten sections. Section 7.2 briefly describes the psychological model of knowledge (a science representation) that underpins the syllabus design application of TAKD. Section 7.3 details the initial Task Analysis (TA) stage of TAKD which produces the first acquisition representation. Section 7.4 details the development of further acquisition representations that allow the low level acquisition representation described in Section 7.3 to be generalised (i.e. represented at a higher level) such that similar task components are represented in a common form (Generic Actions and Generic Objects). Section 7.5 describes how these acquisition representations were further developed so that Generic Actions and Generic Objects come to form the lexical entries to a Knowledge Representation Grammar (KRG). The final acquisition representation is described in which the originally analysed tasks are redescribed by rewriting the steps in each task as Generic Actions and Objects as KRG sentences. Section 7.6 details how the initial syllabus was created from the derived KRG sentences and is the first application representation. The section also describes how the syllabus was represented in a form understandable to its users and how specific training courses can be designed from the syllabus. Section 7.7 describes how the syllabus was evaluated with respect to its ability to represent new tasks and its likely capacity to cope with future changes in IT equipment and practice. Section 7.8 discusses the use of TAKD for the development of training courses in other areas. Section 7.9 summarises the complete TAKD method. Section 7.10 considers applications of TAKD other than the design of syllabi and training courses. In particular, its appropriateness for developing user-oriented task models for input to system design in HCI are examined and contrasted with alternative approaches.

7.2   The science representation

The primary requirement of the syllabus was that it should cope with the diversity found across ITeCs due to the different resources available in ITeCs. The solution to this constraint was for the syllabus to specify the knowledge required to perform IT tasks rather than, as is traditional, the tasks themselves. Only at the level of courses designed within ITeCs on the basis of the syllabus would training tasks be specified. Given this solution, it is clearly necessary to have a model of knowledge. Within Long's framework such a psychological model is a science representation. The particular model chosen was developed, in part, for this research project and is consistent with the psychological evidence concerning human memory and skill. It has properties that are capitalised on in the ITeC syllabus application of TAKD; most importantly, the creation of the acquisition representations in TAKD follows a method that is isomorphic with the hypothesised structure and processes of human knowledge, though no claims are made as to the relationship of the contents of the hypothesised human mental architecture and the content of the products of TAKD. The direct structural relationship between the two does allow, however, the TAKD representation to make predictions about human behaviour, particularly with respect to the consequences of teaching style and the likelihood of successful transfer of training.

Diaper (1984) has proposed that psychological models can be represented as existing at a number of different levels of abstraction. He claims that this is analogous to how physics represents the world at different levels (e.g. the mesoscopic, microscopic, molecular, atomic etc.). In both psychology and physics each level of representation is potentially complete in that, in principle, each level is capable of representing everything without recourse to higher or lower levels of generality or abstraction. For example, water at standard temperature and pressure can be described either as a transparent liquid at the mesoscopic level or as water molecules, that possess hydrogen bonding, at the molecular level. Each level of description is independent, though they can be related via a set of mapping rules that in the water example could be crudely portrayed as the "is made of" function. Similarly, behaviour can be represented at a low level, for example, as a motor program (the final psychological representation prior to thought being translated into physical behaviour) that organises and instructs the musculature to press a sequence of keys so as to type a command. The same behaviour could, however, be described at a higher level as deleting a word in an editing task. Thus, in the same way as water can be described either as a liquid or as molecules, so human behaviour can be described at different levels of generality.

Both of the example psychological descriptions of behaviour mentioned above can be thought of as different representations of the knowledge that the operator may possess. A low level representation is always necessary for task performance since the execution of behaviour logically requires the existence of low level representations such as motor programs. A person may already possess an appropriate low level representation, particularly for highly practiced tasks. Alternatively, such a representation may be generated from a higher level, more general representation. For the successful translation of high to low level representations to occur it is hypothesised that intermediate representations will also have to be generated or otherwise be available. A person will fail to perform a task if such high to low level translation fails, even when the person possesses an adequately complete high level representation. For example, a person may be an expert author, and know what s/he wishes to write, but will fail at the authoring task if required to use a novel word processor where the commands are not known. The author will be unable to generate appropriate low level representations to support behaviours such as typing the correct sequence of characters or finding the correct keys. To take a second example, it is often the case that a person may understand the principles of how an internal combustion engine works but be unable to translate this into any useful car maintenance.

It is hypothesised that much lecture-style, classroom teaching involves presenting students with a relatively high level representation which may be understood but cannot be applied in appropriate situations because the student has not learnt how to convert such knowledge into behaviour. In contrast, hands-on training requires at least a low level representation if the student is to perform a task at all. However, in such hands-on exercises it is possible for the student to fail to generalise the specific training experience (i.e. to generate higher level representations of the behaviours involved in executing training tasks) such that the student will not be able to perform other, similar tasks, even though the training tasks may be performed perfectly. It should be pointed out that verbal behaviour is not treated in a manner different from other behaviours, though the task demands are usually radically different. For example, the task requirements of describing or explaining are very different from the task requirements of manually performing a car maintenance task. Such differences, within the psychological model, are captured by the different translation processes required to verbalise high level knowledge or to translate it into manual behaviours.

Ideally, the knowledge that a person possesses should be at the highest level of representation that allows the knowledge to be translated into behaviour. This allows the person to perform the widest range of tasks consistent with their knowledge. The model predicts that the higher the level of representation of knowledge a person possesses the wider the range of tasks the person can perform, except in the case where a high level representation is so incomplete that lower level representations cannot be successfully generated from it. It follows that the higher the level of representation the more intermediate levels of representation intervene between the high level representation and the necessary low level representation required for task performance and thus it is harder to translate very high level representations into adequate behaviour.

To maximise the general applicability of hands-on training a range of tasks needs to be experienced and emphasis placed on trainees perceiving what is common between the experienced tasks at increasingly higher levels of representation. It is predicted by the model that the success of hands-on training with respect to the transferability of the knowledge gained and the range of tasks experienced will be an inverted U-shaped function when the amount of training on each task is controlled. This sort of function is predicted because if two training tasks are very similar in the behaviours they require, then there will be insufficient contrast between them to promote generalisation (i.e. the formation of higher level representations). If two training tasks are highly dissimilar then generalisation will also be hindered as any similarities are likely to be only perceivable by someone who already possesses a relatively complete high level representation. Thus, in hands-on and lecture-style teaching the emphasis should be in opposite directions. The latter needs to provide knowledge at a sufficiently low level that it can be translated into behaviour and should try to provide guidance on how such translations are possible. Hands-on training should encourage trainees to generalise their specific experiences, and be presented in such a way so as to promote the formation of higher level representations. Thus a combination of lecture style and hands-on training should, if properly approached, complement each other and operate to their mutual advantage.

The ITeC syllabus provides a single high level description of what is required to perform IT tasks. It is called a knowledge description on the grounds that it represents a description of task performance requirements. No claim, however, is made that the actual representational format corresponds to any human mental representation. What is claimed is that a human expert in IT will possess, or be able to generate, a mental representation at a similar level of generality. This is equivalent to psychologists representing a hypothesised mental store such as short term memory (STM) as a box with input/output arrows. The cognitive psychologist is not claiming that there is a box in the mind labelled STM but that there is some mental architecture or mental processes that operate as if such a box existed. While the ITeC syllabus represents only a single, high level representation of IT knowledge the TAKD method exploits Diaper's multi-level psychological model of human knowledge representation in the generation of the chosen level of knowledge representation. In particular, the lower level representations of the tasks used for syllabus creation are known and are consistently mapped to the selected level of representation in the syllabus. There is thus an isomorphism in the method employed by TAKD to generate the syllabus and the processes used by a person to perform tasks utilising a similar level of knowledge representation. This has the consequence of giving the syllabus, via its method of creation, predictive capabilities with respect to human behaviour. In particular, the number of syllabus members that a trainee acquires across different tasks not only specifies the amount of knowledge of IT the trainee possesses but also predicts the general range of novel tasks that the trainee is likely to be able to perform (i.e. transfer of training from hands-on tasks previously experienced to new tasks).

The actual level of description employed in the syllabus was determined on pragmatic grounds by employing the following three criteria. First, the level of generality would have to be sufficiently high that the syllabus' specifications were equipment and task independent. Second, the level of generality was sufficiently low that it could provide concrete guidance to course designers and course evaluators such that it could be relatively easily mapped on to IT tasks. Third the total number of syllabus members was appropriate for the length of course. These last two criteria are in fact different expressions of the same criterion and reflect practical considerations concerning the amount of time and effort required for course designers and evaluators to use the syllabus. The number of syllabus members, not surprisingly, is inversely related to the level of description employed by the syllabus (i.e. a syllabus that represents knowledge at a high level will have fewer members than a syllabus that uses a lower level of representation). It also follows that two courses that are supposed to cover the same material, but in radically different times, should have syllabi specified at differing levels of generality, with the longer course having a larger syllabus at a lower level of representation than the equivalent shorter course. There must be a lower boundary on the level of representation used in syllabi, and therefore their length, as clearly a syllabus containing thousands of members will be unusable because of the demands such a syllabus would make on course designers, evaluators and trainers.

The MSC research remit in fact required that two syllabi be produced that covered the same material. One was to be an introductory, two week course (described in Johnson, Diaper and Long, 1984) and the other was to be the one year ITeC syllabus that this chapter describes. Using the same data from a set of analysed IT tasks, and thus describing the same material, the syllabus for the short course contains 22 members whereas the syllabus for the one year course contains 218 members.

7.3　Task analysis

There are now many different forms of Task Analysis (TA). Wilson, Barnard and Maclean (1986) review eleven that might all be used in HCI research. The selection of an appropriate TA technique depends on the purpose of the TA. This will determine what features of the task will be studied. TA techniques may have a purely descriptive output, as in the case of traditional work study, or may have predictive capabilities, as is claimed by Barnard (1987) for Cognitive Task Analysis (CTA). Annett et al's (1971) Hierarchical Task Analysis (HTA) focuses on the goals, sub-goals and procedures of tasks so as to identify those task components on which people require training, whereas a technique such as Payne and Green's (1986) Task Action Grammar (TAG) focuses on the person, rather than the task, to produce a competence model of the task performer. Moran's (1981) Command Language Grammar (CLG) lies somewhere between HTA and TAG as it is concerned with apportioning tasks between people and machines. TAKD, like TAG, focuses on the performance of the person because its purpose is to identify the human knowledge requirements necessary for successful task performance. Within the work described here, TAKD focuses on producing knowledge specifications from expert task performers so as to generate a high level "ideal user" knowledge model.

It is usual in most TA work to start with some initial, often informal, data collection to map out the general area. Such preliminary stages of the research involved searching the available literature and documentation on IT courses, classifying the contents of existing IT syllabi, and carrying out a series of unstructured interviews with trainers, course designers, examiners and employers. The information collected provided insights into current beliefs concerning what should be taught in IT courses. It was clear from the interviews that the number of opinions on what and how IT should be taught was approximately equal to the number of people interviewed (this highlights one major problem of syllabus design based on subjective, albeit expert, opinion). The analysis of extant syllabi involved first representing each syllabus in a hierarchical form and then amalgamating these so as to generate three hierarchies, one for each of the three areas of microelectronics, computing and automated office applications. Nine, sixteen and fifteen syllabi, or parts of syllabi, respectively were analysed in these areas. These syllabi were selected such that while very advanced material was excluded (e.g. from the final year of a 3 or 4 year industrial training course, or degree level material) the syllabi selected contained more material than even the best ITeC trainee might be expected to master in a one year training programme. This classification provided an extremely useful starting point for the major TA aspect of the research as it indicated the type of material considered important to teach trainees. It was also helpful in guiding some aspects of the subsequent analysis method. The full classification is documented in Long, Johnson and Diaper (1983).

The formal TA involved an initial analysis of 33 tasks performed in commerce, industry, and, in a few cases, in ITeCs. The range of tasks was roughly guided by the MSC's decision that training would cover the three areas of microelectronics, computing and automated office applications. Tasks were analysed that were roughly appropriate to the length of the ITeC training course (one year) and the initial ability of trainees (relatively low as they were generally 16 year olds who had left school with few academic qualifications). The lowest level of description of tasks (the data recorded on site) was partly determined by what was practical (video recording was not practical in this case) and fulfilled a criterion that this lowest level of task description was lower than any subsequent analysis of the data required, (pilot studies had helped to establish this). The data collection concentrated exclusively on the behaviour of the task performer as it was this person's knowledge that was to be identified, rather than that of someone more distant, but associated with the task, such as a supervisor.

The researchers would arrive at a selected site and would analyse whatever suitable tasks were being performed at the time. Rough checks were continually made to ensure that tasks were sampled from all three ITeC areas of training across all the sites visited. As with any sampling exercise, the larger the sample the more reliable the resulting analysis, and conversely, as the sample size increases the effect of adding more observations reduces each observation's impact on the overall results. That 33 tasks were sampled was as much determined by the available time as by any other criterion, though this size sample was believed to be sufficiently large that the impact of any additional task would be relatively small. A subsequent evaluation, described later in this chapter, in general supports this claim.

All the tasks analysed were performed by people who were expert at the task. The goal of the analysis was to produce a description of the knowledge a competent person required for task performance. This can be contrasted with an alternative approach of analysing how trainees performed, or failed to perform, tasks, which would have been appropriate for identifying how trainees learn, rather than what it is they should learn. The data collection involved a researcher making detailed notes of the behaviours performed while sitting beside the expert task performer. The task performer was also required to produce a concurrent verbal protocol which was recorded on an audio cassette recorder. Occasionally the researcher would also ask the task performer what it was they were doing, particularly in those cases where the performers were reticent. The taking of verbal protocols undoubtedly slowed down task performance but, given that the task performers were experts at the task, it probably had little impact on the specific actions or sequence of actions performed. This data collection stage of TAKD is similar to nearly all other forms of TA, though the particular aspects of task performance that are recorded differ. The product in this case was a task description of the sequence of behaviours performed and the objects to which behaviour was directed. These task descriptions form the first acquisition representation in that they each describe a real world task. Only some aspects of the task have been captured (principally the behaviours of the task performer) and this selection has been determined by the final purpose of the research to produce a syllabus, (i.e. by the application representation).

7.4   Generalised acquisition representations

Little can be done with the complete, first acquisition representations because they contain a range of disparate information about objects, behaviours and sequences of behaviours with no common representational format. The second acquisition representation concentrates on only some aspects of the information in these first acquisition representations. In particular, sequential information is ignored. This second acquisition representation consists of two lists that detail, independently, all the particular behaviours performed in the tasks and all the objects to which these behaviours were directed. These two lists of Specific Actions and Specific Objects fulfill a criterion of being exhaustive with respect to the tasks analysed. Furthermore, these two lists have collapsed the information from all of the analysed tasks.

The third acquisition representation also consists of two lists: Generic Actions and Generic Objects. These two lists are created from the lists of Specific Actions and Specific Objects and represent a higher level, more general, description of the lists and thus also of the analysed tasks. An important requirement (the exhaustiveness criterion) is that every member of the Specific Action and Specific Object lists must be an exemplar of at least one Generic Action or Generic Object list respectively. A degree of trial and error was involved in the initial construction of these two generic lists. Importantly, the generification process has as its inputs a low level description, the Specific Action and Specific Object lists and also a very high level description. This latter description was provided by the analysis performed on the same tasks for constructing the previously mentioned short, two week, IT training course which had used an abbreviated form of the TAKD method described here. In essence, the researchers cycled between the high and low level descriptions looking for a set of Generic Actions and Generic Objects that satisfied the exhaustiveness criterion and such that the Generic Actions and Generic Objects were all at approximately the same level of description. Subsequent research on the analysis of student messaging tasks (Diaper, 1988) and for system design (Johnson, Johnson and Russell, 1987a) has concentrated on formalising this generification process. It is suggested that a descriptive hierarchy can be constructed that will link the highest level description, which can describe all the tasks, with the lowest level provided by the Specific Action and Specific Object lists. In the construction of such a task hierarchy the exhaustiveness criterion is maintained at every level of the hierarchy, which must be a true hierarchy without recursions, such that any particular task may be described by a single pathway. There are numerous subjective decisions to be made in the construction of such hierarchies, for example, the specification of the meaning of each node and the alignment of levels. Once constructed however, generic entities may be specified by slicing through the hierarchy at one level.

There can be no absolutely correct specification of generic entities but this is probably only a minor weakness in the TAKD method and such subjectivity occurs, though is usually unadmitted, in other TA techniques. TAG, for example, does not describe how to determine a "simple task". There are not strict criteria for choosing the level of specification of the Generic Actions and Generic Objects, though there is a fairly obvious heuristic regarding the size of the lists, which is, of course, directly related to the level of description, as the lists need to be of manageable length. A list of 100 Generic Actions, for example, is too long (i.e. at too low a level of description), in nearly every imaginable case, as it would take too long to learn, or to apply, by course designers. For the 33 tasks analysed here, the Generic Action list contained 11 members and the Generic Object list contained 24. For ease of use, the Generic Object list is further divided into three categories: physical objects; physical or informational objects; and informational objects. Physical Generic Objects have Specific Object instances such as hand-tools, electrical components or printed circuit boards (PCBs) and can all be physically manipulated. Informational Generic Objects have specific instances such as text or graphics and operations on them are non-physical (i.e. one can read and alter text but cannot manipulate it in the same way as a physical object – in general physical objects can be directly manipulated whereas informational objects can only be changed via the use of a tool such as a pencil, eraser or a QWERTY keyboard). There is also a class of objects that may be either informational or physical depending on the nature of the task. Thus, in automated office and computing applications a computer running software is an informational object whereas when being constructed or repaired for example, it may be a physical object. In the Generic Object list presented in Appendix II, the first six members are physical objects, the following twelve may be either physical or informational and the final six are purely informational.

There are in many cases a number of different specific meanings to the Generic Actions and Generic Objects and examples are provided in Appendices III and IV. These are discussed later in section 7, which considers the evaluation of the syllabus.

The third acquisition representation of Generic Actions and Generic Objects thus constitutes a high level description of the behaviour and the associated objects in the 33 analysed tasks. Still missing from this acquisition representation is firstly the relationship between the generic entities and secondly the sequencing of behaviours. As it turns out, this latter aspect of the tasks was not germane to the creation of the syllabus and is therefore only cursorily dealt with later. In other applications, it is likely, however, to be of crucial importance. The next section deals with the problem of relating Generic Actions and Generic Objects.

7.5 The knowledge representation grammar

The fourth acquisition representation involves the production of a means of relating Generic Actions and Generic Objects. The TA concentrated on the behaviour of the expert task performers and this behaviour is now captured in the list of Generic Actions. It was therefore decided that Generic Actions would be the central high level knowledge elements of interest and that in each case, where a behaviour was performed in a task, a sentence relating Generic Actions to Generic Objects would emphasise the behaviour. These descriptive sentences would contain only a single Generic Action. The construction of the fourth acquisition representation, then began with the development of a Knowledge Representation Grammar (KRG). An early version of the KRG was based on the pairing of a Generic

Action with a Generic Object. While this had the advantage of simplicity
in that the only grammatical rule was that each legal sentence contained
one Generic Action and one Generic Object, it had a number of disadvantages. First, a large number of such simple KRG sentences are needed to
describe even a single step in a task. Second, the description is inadequate
unless a set of such simple KRG sentences is considered, since the meaning
of a single Generic Action-Object pair is modified in the context of other
pairs. In particular, there was a problem with the specification of the location of Generic Objects that were acted upon. An advantage of this simple
KRG is that the grammar is entirely independent of the lexical entries (the
Generic Actions and Generic Objects). The sheer number of such possible pairs, however, made it almost certainly unusable by course designers
and evaluators and it was thus rejected as unsatisfactory. The criterion of
lexical independence was relaxed in the subsequent development of the final KRG, described below, for the Generic Actions but was maintained for
the Generic Objects. Thus, while each Generic Action developed a slightly
different syntactic form, these differences are minor and consistent.

While the precise form of a KRG will vary with the application, the general form is likely to be similar. This has been found to be the case for
describing messaging tasks (Johnson, 1985; and Diaper, 1988). For those
interested in applying TAKD for other purposes, the KRG described below
can be treated as an example of how such grammars can be constructed.
There will, of course, be great differences in the knowledge elements, but
less difference in the grammar that relates such elements. The KRG developed was a modification of the original Generic Action-Object pairing
grammar which provides a good basis from which to start the production
of more sophisticated KRGs such as the one described below.

Each KRG sentence used in the final redescription of the analysed tasks
consists of a Generic Action that operates on one or more Generic Object
phrases. All legal KRG sentences contain only a single Generic Action,
though they may contain between one and three Generic Object phrases,
depending on the particular Generic Action. A Generic Object phrase can
consist of up to two Generic Objects. First, there is the main Generic Object
that is acted upon. Second, there is an optional Generic Object which
specifies, when relevant, the location of the main Generic Object. Such
locative Generic Objects are important as the behaviour directed towards
an object such as an electrical component, for example, is usually different
when such a component is mounted on a PCB than when it is in a box of
loose components. The Generic Action operates on each Generic Object
phrase in a KRG sentence. Furthermore, each Generic Action possesses a
specified number of possible Generic Object phrases such that these phrases

have a consistent relationship to the Generic Action. For example, all KRG sentences that contain the Generic Action "Insert" possess three Generic Object phrases such that all "Insert" KRG sentences potentially specify the following:

> INSERT an object A at, in or on a location B
> INSERT to an object C at, in or on a location D
> INSERT with an object E at, in or on a location F

Where objects A, C and E are the main Generic Objects of each Generic Object phrase and B, D and F are the optional, locative Generic Objects in each Generic Object phrase. Generic Object phrases may be unspecified in a KRG sentence depending on the context, though the order of Generic Object phrases is invariant. As the description always focuses on the task performer, s/he is taken as given and never specified as a location (i.e. there is no locative to specify when a screw-driver, for example, is in someone's hand, though a locative would be used if the tool was in a storage rack).

For example, the act of inserting (A1) the wires (O4) of a capacitor (O6) on to a PCB (O5) that is held in a clamp (O1) using a pair of pliers (O1) can be represented as:

> INSERT (A1) a connector (O4) on an electrical component (O6)
> INSERT (A1) on an electrical object (O5) in a hand tool (O1)
> INSERT (A1) with a hand tool (O1)

Where the indicators in parentheses identify the various Generic Actions and Generic Objects (see Appendices I and II). It is possible for course designers and other syllabus users rapidly to learn the two lists and their numeric indicators and once mastered this allows KRG sentences to be quickly written or read in a short-hand form. In this form, each Generic Object phrase is specified between the slashes. The locatives, where appropriate, are in parentheses. The above sentence would be written as:

A1/O4(O6)/O5(O1)/O1/

As a fairly sophisticated language, it is possible to write different, synonymous KRG sentences. A convention has been adopted in such cases by following a rule that prefers KRG sentences that have the maximum number of Generic Object slots filled. This rule was adequate in all cases of possible synonymity found in this work. It has the consequence, however, of occasionally causing the order of Generic Object phrases in KRG sentences to appear slightly odd to native English speakers. The most obvious case of this is with Generic Actions such as "insert" when used in the context of inserting text on a word processor, where an approximate English description of the KRG sentence

A1/O23(O12)/O19(O7)/O23(O11)/

INSERT (A1) a textual component (O23)
                on a textual output device (O12)
INSERT (A1) to an implicit textual object (019)
                on a computer (O7)
INSERT (A1) with a textual component (O23)
                on a textual input device (O11)

might be translated as: a letter (O23) is displayed (A1) on a VDU (O12) which has been sent (A1) to a word processor program (O19) running on a microcomputer (O7) by typing (A1) a character (O23) on a QWERTY keyboard (O11). The more natural English sentence would have placed the act of typing prior to the appearance of the character on the VDU, though from a subjective point of view the two events are simultaneous (or should be). The researchers quite rapidly learnt both the meaning of Generic Actions and Generic Objects and the KRG such that dialogues about the research could be conducted using only the short-hand form of KRG sentences described above. This learnability is mentioned again later as an important aspect of the application of the syllabus for course designers and evaluators.

The fourth, and final acquisition representation was produced by returning to the first acquisition representation (i.e. the original task analyses) and taking each task step and redescribing it using one or more KRG sentences. Thus, the initial collapsing of the tasks is reversed and the fourth acquisition representation provides a high level description of each task in a uniform representational format. As all of the Specific Actions and Specific Objects, which were themselves an exhaustive list of all such actions and objects in the analysed tasks, are also exhaustively represented at the generic level of description, it must be the case that all the task steps are redescribable as KRG sentences. This was found to be the case.

While this constituted the final acquisition representation here, in other applications of TAKD a fifth representation may be appropriate if sequence information is required. Such information may be captured by analysing the tasks represented by KRG sentences for common occurrences of patterns of KRG sentences.

7.6   Syllabus creation and course design

Up to this point, the TAKD method has provided only acquisition representations, that is, it has been involved with the description of real world tasks. The syllabus that was created is the first application representation in that it could, in theory, be mapped back on to the real world of IT tasks. The first version of the syllabus, however, should not be confused with subsequent, post-evaluative versions and the final version. The first application representation consists entirely of KRG sentences and, while comprehensible to the researchers, would have been totally unsuitable for those who were to use the syllabus. This first application representation was created by listing each KRG sentenced required by each analysed task, ignoring repetitions of KRG sentences within tasks. The syllabus then consisted of the total set of KRG sentences found to be necessary to describe all the analysed tasks. While there are a very large number of possible KRG sentences, many of the total possible set are meaningless and this initial form of the syllabus contained 218 KRG sentences. This was considered satisfactory as it was fully representative and judged to be a manageable set. If the set had contained many hundreds, or over a thousand, KRG sentences then this would certainly have been unmanageable and would have implied that too low a level of description had been employed in the construction of Generic Actions and Generic Objects, or that the KRG was itself insufficiently complex in its ability to combine Generic Actions and Generic Objects (the problem with the early Generic Action-Object pair grammar). Similar concerns would have been expressed if less than about 50 KRG sentences had been required as this would have implied too high a level of description of either the Generic Actions, Generic Objects or the KRG. It is worth pointing out that such criteria depend on the final application and user population of the syllabus.

The second application representation added structure to the list of KRG sentences described above. A number of organising principles were considered. Originally the syllabus was organised into groups of KRG sentences with those that were found to be necessary in tasks from all three of the areas of ITeC training (microelectronics, computing and automated office applications) being given precedence over those required in only two of the three areas, and still less precedence being given to sentences required in only one of the three training areas. Within each of the seven possible groups, the KRG sentences were ranked in order of the number of tasks that had required each sentence, with the most frequent being listed first. This grouping was abandoned, however, as it turned out to be too difficult for the MSC and for a small, sample set of course designers to use. A simpler organising principle was finally adopted where KRG sentences were just ranked according to their frequency of occurrence across tasks, with the most frequent appearing earlier in the syllabus than the less frequent. The frequency of occurrence across tasks ranged from 72% to 3%. A frequency of 3% implies that a particular configuration of Generic Actions and Generic Objects (a KRG sentence) was only found to be required in a single task analysed. Of the 218 KRG sentences in this syllabus, 100 (46%) had a frequency of occurrence greater than 3%.

There are a number of slightly different methods that can be used to design a specific training course from the syllabus. They all, however, require that a range of possible hands-on training tasks be selected and KRG sentences assigned to them. KRG sentence assignment can either be effected without referring to the syllabus or, more easily, by going through the syllabus, identifying and marking those syllabus members that are required for each task's performance. Only now can the tasks that might be used in a training course be selected on the basis of the number and range of syllabus members (KRG sentences) covered by the set of tasks. Over the course of a year, the trainees should be exposed to every syllabus member many times and the frequency of the exposure should approximate that found in the syllabus. Those trainees who cannot master, in a year, the whole syllabus, should at least be familiar with the higher frequency syllabus members, as these are essential to the widest range of IT tasks. The same method can be used for the evaluation of complete training courses.

In-house trainee assessment is possible by calculating a profile of those syllabus members a trainee successfully uses in performing the hands-on tasks experienced. Given that the necessary syllabus members are known for each task this could be fairly simple if a copy of the syllabus is used as a record of each trainee's progress. There is, of course, considerable scope for the automation of both the course design process and the maintaining of trainee records. Formal assessment of trainees is facilitated by the syllabus as novel tasks used to test trainee performance may be selected/designed on the basis of the KRG sentences required for such task's performance.

7.7   Evaluation and the final product

The syllabus was formally evaluated with respect to two different properties: (i) its completeness; and (ii) its ability to cope with future technological change (New Information Technology – NIT). The completeness of a syllabus refers to its ability to describe all the possible tasks within its scope. No syllabus could ever be demonstrated to be 100% complete, but it is useful to know its approximate completeness and, of course, a highly incomplete syllabus will fail to provide proper guidance to its users. With respect to the ITeC syllabus, this has been tested by examining additional tasks that were not used in its creation. An alternative might have been to compare the contents of the ITeC syllabus against other syllabi. This was not done as other syllabi make no formal claims to being complete and there is no obvious selection criteria or method of comparison between syllabi. The ITeC syllabus describes the knowledge required to perform existing tasks in IT. Given the rate of technological change in this field, it is desirable that the syllabus be able to cope with such changes and that there be a mechanism to allow the syllabus to be tested and changed, if it is found to be out of date. Such an evaluation can be performed as a test of completeness, since a syllabus that becomes dated will have become less complete with respect to current IT practice. A method of anticipating change, rather than trying to catch up with change, is demonstrated following a description of the testing of the completeness of the syllabus.

Seven additional tasks were chosen on the same basis as the original set of 33 analysed tasks and treated in an identical manner (i.e. each was analysed and described as an activity list before having KRG sentences assigned to them). In addition, the frequency structuring of the syllabus was checked by correlating the frequency of occurrence of syllabus members against the frequencies found in the additional tasks.

The seven additional tasks required a total of 137 KRG sentences to describe them (range = 17 to 27 different KRG sentences per task). No objects or actions were found in the additional tasks that were not exemplars of the Generic Actions and Generic Objects (i.e. the Generic Action and Generic Object lists were complete). The additional tasks required 58 different KRG sentences. Of these, 72% were syllabus members. The proportion of syllabus member KRG sentences required by each task ranged from 53% to 100%, with a mean of 87%. Of these omissions from the syllabus, 55% came from a single task: constructing a half-wave rectifier. This was the only additional task purely from the area of traditional microelectronics. Of the additional tasks' KRG sentences that were syllabus members, 86% had a frequency of occurrence greater than 3% in the syllabus. This compares with the 46% of syllabus members with frequencies above 3% in the syllabus. The z score from a Kendall Rank Correlation calculated between the frequency of occurrence of syllabus members against their frequency found in the additional tasks was 4.40 and is highly significant, thus indicating that KRG sentences were similarly distributed across the original and additional tasks.

Other than in the domain of microelectronics, the syllabus was judged to be satisfactorily complete. None of the omissions required any unusual application of the KRG and most of them concerned connecting and disconnecting a disc drive which had not been observed in the original 33 tasks.

Microelectronics, which the MSC had defined before this research project, included not only traditional electronic skills such as circuit assembly and testing but also microchip and robotics programming. It appears that the area of traditional microelectronics (i.e. not machine level programming) is specified in the ITeC syllabus at a slightly lower level of generality than the other applications areas. If only those syllabus members that are exclusively involved with traditional microelectronics tasks are considered then the above claim is supported by the fact that 47% of all syllabus members come from such tasks and their highest frequency of occurrence is only 12%. The mean frequency of such microelectronics syllabus members is only half the mean frequency of the complete syllabus (4.5% versus 9.2%). The mean frequency of occurrence of syllabus members across tasks is weighted by the skew caused by this large number of traditional microelectronics related syllabus members such that if they are removed from the syllabus then the mean frequency of occurrence rises to 14.1%.

Those KRG sentences required in the additional tasks that were not syllabus members were added to the syllabus. This requires that all the frequencies in the new syllabus are adjusted as the new syllabus is now based on 40 (33 + 7) tasks. It is possible to continue adding tasks to the syllabus in this way, though if done frequently, criteria of task selection will be necessary so as to prevent the syllabus becoming biased by the incorporation of a large number of the same type of task.

In the future, if the syllabus is thought to have become out of date, this can be tested by examining additional tasks, particularly those that have become more common in industry, commerce etc. since the syllabus was first created. Anticipating technological change is more difficult and while several different methods were considered (e.g. interviews with experts, reviewing government sponsored research and development such as the Alvey programme) only one method is described here. Given the uncertainty of the product development and that many products that are marketed do not become common, then testing the syllabus' ability to cope with new IT should not lead to altering the syllabus, though it can alert syllabus reviewers to those areas where the syllabus is likely to become out of date.

The method of testing the sylabus' ability to handle new IT was to select a number of magazines in the three training areas of ITeCs. The magazines' advertisements were used as a data-base of future IT applications on the grounds that they are attempting to establish products so that they do become common in IT. Each advertisement was examined with respect to the specific objects advertised and classified in one of three categories: (i) objects that had been encountered in the 40 tasks used for syllabus creation and the completeness evaluation; (ii) objects that were novel examples of the existing Generic Objects; and (iii) objects that could not be accommodated by the existing Generic Objects. There is, of course, some subjectivity involved in this classification as in many cases it is difficult to differentiate the manufacturers' claims of novelty from truly new IT objects. The criteria were applied such that if there was any doubt, advertised objects were placed in a higher rather than a lower category (e.g. category (ii) rather than category (i)).

All advertised objects considered could be accommodated by the existing set of generic objects (i.e. category (iii) was empty) and only 7 of the 24 generic objects were needed to describe category (ii), of which 27 new examples were found. To demonstrate that the syllabus can cope with these new examples, imaginary KRG sentences were constructed for these objects and the similarity of such KRG sentences can be noted by substituting a Specific Object that was used in the 40 tasks for the novel object (e.g. the use of a crimping tool (O1) can be expected to be described by the KRG sentences that were previously used to describe the use of a pair of pliers, which had been encountered before). Although no difficulty was experienced in this construction, this does not preclude the possibility that some application of these objects may not be describable by an existing KRG sentence in the current syllabus.

The third and final application representation could now be constructed. Up to this point in the project, the syllabus had only been described by a list of KRG sentences. For it to be understandable and usable by course designers, trainers, course evaluators and the assessors of trainees it needed to be explained and exemplified. This was done in consultation with the MSC and draft copies of the syllabus were circulated to a small set of ITeC managers for them to evaluate. The research project had planned to evaluate the application of the syllabus in ITeCs but due to a number of factors, outside the control of the researchers, this was not possible in the time available. However, the final form of the syllabus (examples taken from the complete syllabus are presented in the appendix V) was judged by those outside the project to be understandable, though it is recognised that this is no substitute for field trials. The syllabus, along with the prose descriptions of the Generic Actions and Generic Objects, constitutes the final application representation. All that was missing were the instructions on its use, the production of which was never finally agreed upon with the MSC. In the syllabus, each member is presented as a KRG sentence with its associated frequency of occurrence. The Generic Actions and Generic Objects were then presented in full, followed by one or more examples of the use of the KRG sentence. These examples are illustrative and were not meant to be exhaustive. The examples maintain the KRG ordering and so provide an easy mapping between the KRG sentences and the prose exemplification of the sentences. The syllabus was structured by the frequency of occurrence of KRG sentences found across the analysed tasks with the highest frequency syllabus members appearing earlier in the syllabus. Syllabus members of equal frequency were arbitrarily ordered by their Generic Action number. This indexing allows syllabus members to be found relatively easily. It was recognised that for some applications of the syllabus this might not be the ideal structure, but in the absence of field trials no others were produced.

7.8   Generalising the research

This section considers how the ITeC syllabus research may be generalised to other training areas. The research produced a method for deriving the content of training courses by an empirical TA and the exploitation of a knowledge model. The syllabus identifies the knowledge that experts require to carry out a task. This use of TA in training research is not itself unique. However, in contrast to other approaches (e.g. Annett et al, 1971; Fleishman and Quaintance, 1984), which identify where training is required rather than what should be learnt, the ITeC research project allows the final specification of the content of training courses from a syllabus that details the knowledge trainees need to acquire.

The criteria that the research project identified and met are not particular to ITeCs or IT training. One assumption that was not questioned in the research, however, was that the style of training was to be hands-on, as ITeCs were set up to provide this. This has the consequence of constraining the type of knowledge that the research sought to capture in that the knowledge identified was confined to that necessary for the performance of IT tasks. The research did not consider such aspects of training as providing explanations, for example, identifying the knowledge used to answer such questions as "Why should I use a heat sink here?" or "What is the difference between a non-recursive and a recursive procedure?". Such knowledge was beyond the research remit of this project. Before the TAKD method is applied more broadly we expect that some effort would need to be made to incorporate knowledge requirements that tend not to be transmitted by hands-on training.

Given such limitations, TAKD is still directly and immediately applicable to a wide range of training problems where hands-on experience is the principal learning method. The methods described in this chapter can be used on a wide range of non-IT topics, particularly where general, transferable training in a topic is required and where there is a wide range of possible tasks. It should therefore be easy to apply TAKD to such topics as carpentry, metal work, plumbing, building construction and some aspects of the retail trade. Obviously the approach can be extended to non-IT office tasks such as the use of manual filing systems (e.g. Kalamazoo systems) and to devices not explicitly covered in the ITeC syllabus such as telephone switchboards, facsimiles and photocopiers, manual typewriters, dictaphones, etc. It is possible, but less clear that training syllabi can be developed from TAKD for tasks that have a large cognitive component (e.g. managerial training). However, there seem to be very few tasks that, while cognitive, do not also have a behavioural output. Tasks such as managerial decision making or financial planning usually involve behaviours associated with the acquisition and reorganisation of information and often trial decisions or solutions are produced before the final outcome. Such behaviours are open to TA and may form the basis for inferences about the knowledge required to carry out such tasks. Many of the tasks analysed using TAKD in the syllabus research in the areas of computing and the automated office had a large cognitive component of this type and this is reflected in the frequent use of such Generic Actions as "perceive", "search", "choose", etc. However, TAKD has not been applied to the design of lecture-style teaching programmes and it is likely that at least some modification will be necessary before the version of the TAKD method described in this chapter will adequately handle such training.

7.9   Summary of the TAKD method

To develop a training scheme in an area that involves a number of different tasks, that all possess a reasonably large component of observable behaviour, the following steps should be followed:

1. A representative set of example tasks are selected. Ideally there should be more than one example of each type of task, though this will depend on what is practical and on the range of possible tasks.

2. Each task, performed by an expert, is recorded and then described as a series of task steps. It is important to ensure that the level of description is sufficiently low that all important task details are captured. Observation should centre on the behaviour of the expert.

3. From the set of task descriptions two lists containing all the Specific Actions and all the Specific Objects are produced. This is the lowest level of analysis used in TAKD.

4. Two lists of Generic Actions and Generic Objects are then produced from the lists of Specific Actions and Specific Objects. This is best done by producing a very high level description and generating a hierarchy of

generic entities by working both top-down and bottom-up. It is important that all Specific Actions and Specific Objects are exemplars of at least one Generic Action and Generic Object at every level in this hierarchy. The level of Generic Actions and Generic Objects finally chosen for syllabus specification should contain a manageable number of entities. The actual number of Generic Actions and Generic Objects will vary across applications, and with the duration of the final course. As far as possible, the level of description should be the same both within and across these lists.

5. A knowledge Representation Grammar (KRG) now needs to be constructed. The form of the KRG is not prescribed in TAKD. It should, however, emphasise the behaviour of the task performer and thus the general form of a Generic Action operating on one or more Generic Objects or phrases should not be replaced unless there are clear needs for an alternative.

6. KRG sentences can now be used to redescribe the tasks originally subjected to analysis by assigning one or more KRG sentences to each task step. As these are the same task descriptions from which the Generic Actions and Generic Objects were derived, it must be possible to describe completely all the tasks as KRG sentences. If this is not possible, then it is likely that either a Specific Action or Specific Object has been omitted from these lists, or that the criterion has not been met that every Specific Action and Specific Object must be an instance of at least one Generic Action or Generic Object.

7. A syllabus can now be constructed. There are many possible ways of organising a syllabus and presenting it. This will be strongly determined by the intended uses of the syllabus and by its target audience. Serious consideration should be given to the syllabus' presentation and its methods of application.

8. In most cases the syllabus should be evaluated, at least for its completeness, by analysing a new set of tasks. The KRG sentences found to be required for such new tasks can be compared to those already contained within the syllabus with particular attention being paid to omissions from the syllabus. A syllabus that is found to be excessively incomplete by such testing has probably been designed from an insufficient number or range of tasks. In such cases, more tasks should be analysed. The presentational format of the syllabus should, of course, be extensively evaluated during its early use and anything that is unclear should be modified. Often this is likely to take the form of extending the range of examples supplied with the syllabus. Such presentational modification will not change, however,

the actual content of the syllabus.

7.10   TAKD for system design

This final section summarises how TAKD is currently being developed in applications not concerned with training. In particular, TAKD's appropriateness for developing user-oriented task models as input to system design in HCI is now being explored. Some work, already published, briefly describes such HCI applications of TAKD (Johnson, Diaper and Long, 1985a; 1985b; and Johnson, 1985) and a review of HCI research and TA can be found in Johnson and Johnson (1988).

One of the main concerns of HCI is how to bring the design of computer systems, and user interfaces in particular, more into line with users' needs and thus to make computers more effective tools that are safer and easier to use. There is little dispute about the necessity of satisfying this goal and most designers now recognise the need for some form of input from the human sciences. The human sciences can make two contributions to design: first, in terms of evaluation; and second, in the identification of user requirements (e.g. Card, Moran and Newell, 1983). This latter application is less common than the former as it is more difficult to see how analytic techniques such as TA can contribute to design in the absence of even a design specification. Below we consider how TAKD could contribute to user requirement specification. A common assumption of good design is that there is a match between the users' existing knowledge of their tasks and the knowledge required to carry out the same tasks on a new system. Such maintenance of user knowledge must exist at a high level of generality that supports existing, relatively low levels of knowledge representation as low level knowledge associated with existing tasks is likely to be redundant in the new tasks. New systems should not be designed to mimic old systems, or manual systems, at the lowest level, as this leads to inefficient use of the available technology. This has occurred in cases where a system slavishly mimics double entry book-keeping, in accountancy applications, for example. New systems should support users' existing goals, and the general way they go about achieving them and preferably extend the scope of tasks which users are able to perform. TAKD can be seen to have a potential a role in the design process in establishing the minimum requirements of a replacement system and identifying critical features of user's existing knowledge of the task domain.

In "top-down" or "decompositional" approaches to system design where design proceeds from a high level, general model, through a more specific model to an implemented prototype, the designer constructs a logical model of the application domain. Various approaches to structured design and modelling have been developed (e.g. Jackson Structured Design (Cameron, 1984)). Such approaches are well developed and clearly necessary but they fail to capture how individuals perform tasks. For example, models developed by these approaches would characterise all the entities and information flows in a financial system, such as the London Stock Exchange, but would not describe how such tasks as share price monitoring, or buying and selling, are performed. This is a serious omission from any model used to design interactive systems as the logical model does not reflect users' tasks, goals or the means by which they achieve goal satisfaction. Task analysis methods such as TAKD, however, focus on what it is that users do in tasks so that the activities or action sequences and their structure are the primary interest of the description, unlike conventional analysis methods such Jackson's.

As the design of a replacement system will change some aspects of a task, particularly at the low level, TAKD would appear to have a considerable advantage over many other TA techniques in that it has been explicitly designed to describe tasks in a manner that is implementation independent. Despite the current popularity of "desk-top metaphors", these provide an example of the possible confusion that can be caused by implementing a low level description of objects at the interface, rather than using higher level generic objects, as real desks do not have pull down menus, "icons", "mice" nor "windows", which in most offices are glazed holes in walls (Storrs, 1986).

It should also be realised that system designers are used to working with precise, and often formal, specifications. Thus, the output of any analysis technique (the application representation) should also be presented in a manner that is readily understandable and usable by designers. The KRG representation, we claim, could be be readily converted into such an application representation for engineers. In fact, there is a similarity between the approach adopted by TAKD and the object oriented programming approach.

TAKD can also be independent of any one instance of a task, which is probably important in system design, as it is rare that a system is designed for only carrying out one task by a single, homogeneous set of users. Three criteria can thus be identified as being probably necessary for a technique used for requirement specification: (i) it should be independent of the particular technological aspects of a task; (ii) it should be independent of individual instances of the task; and (iii) it should provide a precise output description. TAKD appears to be able to fulfill these criteria, which are similar to the syllabus design criteria described earlier (Section 7.1).

Initial attempts to apply TAKD to system design have been encouraging (Johnson, Diaper and Long, 1985b; Johnson, 1985). More work on this application of TAKD is necessary to determine TAKD's scope and limitations. For example, in the design of a computer system to support group communication the important aspects of the tasks appear to be the social aspects of communication and the knowledge each participant has of this (Buckley and Johnson, 1987). One weakness of TAKD already mentioned is that in the research described here no attempt was made to capture structural information about tasks (i.e. the relationship between KRG sentences). Certainly in the Buckley and Johnson work this was required because the structural relationships determined the relationship between tasks and sub-tasks. Recently an attempt has been made to use a frame based representation to overcome this omission (Keane and Johnson, 1987) by including macro-actions such as "cause" and "enable" that link sub-ordinate frames together.

In conclusion, the MSC syllabus research project typifies the approach necessary for applying cognitive Ergonomics to complex real world problems. This chapter illustrates the utility of Long's framework in forcing the researcher to be clear as to the nature of the products the research and analysis produces (acquisition and application representations) and how these are related to both the real world and to scientific knowledge. There is no doubt that the research reported here produced a syllabus that more than met the initial requirement specifications and that the method of TAKD is suitable for application in other areas of training and, with extensions, to other applications such as system design in the field of HCI.

Acknowledgements

This work was funded by the Manpower Services Commission. The views expressed here are not necessarily those of the funding body. We would like to thank the MSC, the industrial and commercial organisations, the ITeCs and all the trainers and trainees who helped us with our research.

The order of authors in the following references is alphabetical and reprints may be obtained from either author.

References

Annett,J., Duncan, K., Stammers, R. & Gray (1971). Task Analysis. *Training Information* No. 6. HMSO London.

Barnard, P. (1987). Cognitive Resources and the Learning of Human-Computer Dialogues. In Carroll, J. (Ed.). *Interfacing Thought: Cognitive Aspects of Human-Computer Interaction*. MIT Press: Mass.

Buckley, P. & Johnson, P. (1987). Analysis of Communication Tasks for the Design of a Structured Messaging System. In Diaper, D. & Winder, R. (Eds.). *People and Computers III*. Cambridge University Press.

Cameron, J. (1984). *JSP & JSD: The Jackson Approach to Software Development*. IEEE Computer Society Press

Card, S., Moran, T & Newell, A. (1983). *The Psychology of Human-Computer Interaction.* Lawrence Erlbaum Associates: N.J.

Diaper D. (1984). An Approach to IKBS Development Based on a Review of "Conceptual Structures: Information Processing in Mind and Machine" by J.F. Sowa. *Behaviour and Information Technology,* **3**. 249-255.

Diaper, D. (1988). Task Analysis for Knowledge Descriptions: Building a Task Descriptive Hierarchy. In Megaw, E. (Ed.) *Contemporary Ergonomics 1988.* Taylor and Francis: London.

Fleishman, E & Quaintance, M. (1984). *Taxonomies of Human Performance: The Description of Human Tasks.* Academic Press: Florida.

Johnson, P. (1985). Towards a Task Model of Messaging: An Example of the Application of TAKD to User Interface Design. In Johnson, P. & Cook, S. (Eds.). *People and Computers: Designing the Interface.* Cambridge University Press.

Johnson, P., Diaper, D. & Long, J. (1984). Syllabi for training in Information Technology. In Megaw, E. (Ed.) *Contemporary Ergonomics 1984.* Taylor and Francis: London.

Johnson, P., Diaper, D. & Long, J. (1985a). Tasks, Skills and Knowledge: Task Analysis for Knowledge Based Descriptions. In Shackel, B (Ed.). *Interact'84 – First IFIP Conference on Human-Computer Interaction.* Elsevier: Holland.

Johnson, P., Diaper, D. & Long, J. (1985b). Task Analysis in Interactive System Design and Evaluation. In Johanssen, J. Macini, C. & Martensson, L. (Eds.). *Design and Evaluation Techniques for Man Machine Systems.* Pergamon Press: Oxford.

Johnson, P. & Johnson, H. (1988). Collecting and Generalising Knowledge Descriptions from Task Analysis Data, *ICL Technical Journal* May, 137-155.

Johnson, P. & Johnson, H. (1988). Practical and Theoretical Aspects of Human Computer Interaction. *Journal of Information Technology* 3, No. 3 147-161.

In Aleksander, I. (Ed.). *The World Year Book of Fifth Generation Computing Research and Development.* Kogan Page: London.

Long, J. (1986). "People and Computers: Designing for Usability" In Harrison, M. & Monk, A. (Eds.) *People and Computers: Designing for Usability.* Cambridge University Press.

Long, J. (1987). The Application and Use of A Framework for Paradigm Models. Position paper at the conference: *The Future of the Psychological Sciences.* British Psychological Society.

Long, J., Johnson, P. & Diaper, D. (1983). *The Twelve Month ITeC Syllabus.* Report to the Manpower Services Commission, Sheffield.

Moran. T. (1981). The Command Language Grammar: Representation of the User Interface in Interactive Computer Systems. *International Journal of Man-Machine Studies,* **15**, 3-50.

Payne, S. and Greene, T. (1986). Task Action Grammars: A Model of the Mental Representation of Task Languages. *Human Computer Interaction,* **2**, 93-133.

Storrs, G. (1986). Of Mice and Men. *British Computer Society Human-Computer Interaction Specialist Group Newsletter,* April. No. 6.

Wilson, M., Barnard, P. and Maclean, A. (1986). Knowledge in Task Analysis for Computer Systems. In Green. T., Hoc, J., Murray, D. & van der Veer, G. (Eds.). *Working with Computers: Theory versus Outcome.* Academic Press: London.

Appendix I: The generic actions and their syntax

The letters U,V,W,X,Y,Z denote where Generic Objects may be located with the KRG Sentence. Generic Object phrases are represented between the slashes and the optional locative Generic Objects are marked in parentheses. An alternative condensed form of the syntax of Generic

Action A7 and A9 are provided which are equivalent to the full form of each given first.

Generic Action No.	Generic Action Name	KRG Sentence Syntax
A1	INSERT	/a U(on a V)/to a W(on a X)/with a Y(on a Z)/
A2	REMOVE	/a U(on a V)/from a W(on a X)/with a Y(on a Z)/
A3	FASTEN	/a U(on a V)/to a W(on a X)/with a Y(on a Z)/
A4	LOOSEN	/a U(on a V)/from a W(on a X)/with a Y(on a Z)/
A5	TRIM	/a U(on a V)/with a W(on a X)/
A6	CHECK	/a U(on a V)/against a W(on a X)/with a Y(on a Z)/
A7	SELECT	/a U(on a V)/from a W(on a V)/
	or	/a U(from a W)/(on a V)
A8	SOLVE	/a U(on a V)/with a W(on a X)/
A9	SEARCH	/a U(on a V)/from a W(on a V)/
	or	/a U(from a W)/(on a V)
A10	PERCEIVE	/a U(on a V)/
A11	WRITE	/a U(on a V)/to a W(on a X)/with a Y(on a Z)/

Appendix II: The generic objects

The following Generic Objects are substitutable in the KRG sentences listed in Appendix I where the letters U,V,W,X,Y,Z appear.

Generic Object No.	Generic Object name
O1	Hand Tools
O2	Power Tools
O3	Non-electrical Components
O4	Connectors
O5	Electrical Objects
O6	Electrical Components
O7	Computers
O8	Test Equipment
O9	Storage Devices
O10	Storage Media
O11	Textual Input Devices
O12	Textual Output Devices
O13	Graphical and Analogue Input Devices
O14	Graphical and Analogue Output Devices
O15	Robots
O16	Controls
O17	Help Facilities
O18	Explicit Textual Objects
O19	Implicit Textual Objects
O20	Graphical and Analogue Objects
O21	System Facilities
O22	Menus
O23	Explicit Textual Components
O24	Cursors

Appendix III: Example description of a generic action from the ITeC 12 month syllabus (version 3)

A1 INSERT

insert/ an A(on a B)/to a C(on a D)/with an E(on an F)/

This is perhaps the most important of all the generic actions. The first object phrase (OP1) specifies the object that is going to be inserted. The second object phrase (OP2) specifies where the object in OP1 is to be inserted. The third object phrase (OP3) is instrumental and thus specifies any tool or object used in the process of insertion.

The most obvious use of INSERT is in the sense of putting a physical object like a capacitor (O6) onto another physical object like a PCB (O5), which can be simply expressed as A1/O6/O5/-/, or putting a floppy disc (O10) into a disc drive (O9) as A1/O10/O9/-/. INSERT here means locate the object in OP1 onto the object specified in OP2. Obviously, a tool (O1) may be used, for example, using a pair of pliers to insert an electrical component on a PCB is expressed as A1/O6/O5/O1/. It can also be used to describe the act of orientating an object as in A1/O5/-/-/ (e.g. orientate a PCB)

A second use of INSERT, similar to the above, because it involves physical Objects, is the act of plugging in a plug (O4) to a socket (O4) (i.e. locate a plug in a socket). In this use of INSERT the locative can be used to specify the object attached to the connector. It should be remembered that all power supplies are covered by generic object O5 (Electrical Objects). Thus the act of plugging a microcomputer (O7) to the mains is expressed as A1/O4(7)/O4(O5)/-/, which has a literal syntactic translation of "Plug a connector on a computer to a connector on a mains supply".

A third use of INSERT is to move, position, turn, push, lock, toggle etc. switches, dials, controls (O16's) or catches or locks like those found on disc drive doors (O3's). Normally only OP1 is filled. The main generic object specifies the object being acted on and and the locative is used to specify the device on which this object is located. Thus, turning on a microcomputer (O7) can be expressed as A1/O16(O7)/-/-/, turning on a VDU as A1/O16(O12)/-/-/ and closing a disc drive door as A1/O3(O9)/-/-/.

A fourth use of INSERT involves informational objects (e.g. text (O18), menus (O22), short character strings or numbers (O23)). The most common use is typing text on a QWERTY keyboard (O11), this text is usually sent to either a program (O19) or some other system facility (O21) in a computer (O7) and either the text or some other response is displayed on a VDU (O12). When INSERT is used in this way OP1 specifies what is inserted (displayed) on the VDU, OP3, being instrumental, specifies the object used as a locative (a QWERTY keyboard for example) and of course what is typed on it is the main generic object of OP3. OP2 specifies where the main generic object in OP3 is sent. It is debatable if it is necessary to specify OP2, but it was done in the ITeC syllabus' KRG. Thus, the case of typing a number on a microcomputer such that it is echoed on the VDU, can be expressed as A1/O23(O12)/O21(O7)/O23(O11)/. Which has a syntactically literal translation of "Display a number on a VDU, send to a system facility in the microcomputer the number that is typed on the keyboard". Similarly, obtaining a menu on a VDU by a single keystroke while running a program would be represented by A1/O22(O12)/O19(O7)/O23(O11)/. The syntax of this use of INSERT seems a little strange at first, but is maintained to allow consistency with INSERT's other applications.

The fifth use of insert is similar to the fourth in that it involves informational objects such as text (O18), graphs or diagrams (O20's), but differs in the level of technology involved. INSERT can mean to write something (e.g. text) on paper (O10) with a pen or pencil (O1) and is thus expressed as A1/O18(O10)/O10/O1/ or A1/O18/O10/O1/. Greater precision can be achieved, if desirable, for example, the act of putting the numbers on the axes of a graph can be represented as A1/O23(O10)/O20(O10)/O1/ and expressed as "Put numbers on paper onto a graph drawn on paper with a pen".

The generic action A11 (WRITE) is used only for creative writing or drawing. Most of the IT tasks analysed did not involve a creative component and this is reflected in the use of the fourth and fifth uses of INSERT, mentioned above, rather than WRITE, in most of the tasks analysed.

With five different uses to INSERT it is hardly surprising that more syllabus members were found to use this generic action than any other. The uses do not, however, overlap with each other because each use of INSERT takes quite different combinations of generic objects. There is therefore little to be gained from dividing INSERT into two or more generic actions at the cost of the common semantic properties of the five uses of INSERT. Synonyms: locate; position; orientate; switch; push; press; toggle; plug in;

display; type; write; draw.

Appendix IV: Example descriptions of generic objects from the ITeC 12 month syllabus (version 3)

O1 hand tools This generic object includes all unpowered tools such as screwdrivers, pliers etc. and any object that is used as a tool (e.g. a tooth pick is a hand tool when used to clean out small scraps of paper jamming a printer). It is worth pointing out that pencils, pens and other writing and drawing equipment are also hand tools. Care must be taken with respect to objects that are Test Equipment (O8) rather than tools. A ruler, for example, might be a tool when used to draw a straight line, but a piece of test equipment when used to measure, and thus test or check, some other object.

Examples: screwdrivers; pliers; wire strippers; chip insertors; clamps or vices; solder; pens; pencils.

O9 storage devices This generic object can refer to any object that stores either physical objects, such as tools, screws or electrical components, or informational objects, such as files, records or programs. The only exceptions are those of Textual Output Devices (O12) and Graphical or Analogue Output Devices (O14). The rationale for separating objects such as printers (O12) from Storage Devices is that the former's output can be read by a human operator whereas this is not the case with computers' backing store. It follows from this that a paper tape punch is a Storage Device and not a Textual Output Device.

Examples: trays; draws; shelves; boxes; soldering iron stands; disc drives; cassette recorders; paper tape punch/reader; magnetic tape readers; FOR-TRAN cards; books and anything on paper.

O11 textual input devices This generic object is usually used to cover all types of keyboard. It also covers such things as telephone base units with dials or buttons. There are some new input devices such as Optical Character Recognisers (OCR) that may, when they are available, also fit into this category. Pointing systems or voice input would not be Textual Input Devices, however, but Graphical or Analogue Input Devices (O13).

Examples: QWERTY keyboards; number pads; microwriter keypads; telephone base units.

O23 explicit textual components This generic object is used to describe all readable short commands and numbers. It is a subset of Explicit Textual Objects but has been separated because there is a considerable difference between reading a prose text and reading a single number or short commands. Many computer commands are symbolic, acronymic or mnemonic and often one or two key strokes can cause complex operations to be performed.

Examples: operating system commands, disc operating commands, system facility commands, menu selection commands.

Appendix V: Example syllabus members from the ITeC 12 month syllabus (version 3)

Syllabus Member	% Freq. of Occurrence

A10/O18(O12)/   69
PERCEIVE: an explicit textual object
         on a textual output device

e.g. read a program displayed on a VDU
e.g. read a document displayed on a VDU
e.g. read a command string displayed on a VDU

A1/O23(O12)/O19(O7)/O23(O11)/   51
INSERT: an explicit textual component
         on a textual output device
         to an implicit textual object
         on a computer
         with an explicit textual component
         on a textual input device

e.g. a short command is displayed on a VDU that has been sent to a word processor program in a computer by typing a command on a QWERTY keyboard
e.g. a number is echoed on a VDU that has been sent to a program in a computer by typing a command on a number pad

A1/O10/O9/-/   48
INSERT: a storage medium
         to a storage device

e.g. put a floppy disc into a disc drive

A6/O18(O12)/-/-/   27
CHECK: an explicit textual object
         on a textual output device

e.g. check a document displayed on a VDU
e.g. check a program displayed on a VDU

A2/O2/O9/-/   12
REMOVE: a power tool
         from a storage device

e.g. take a soldering iron out of its stand

A1/O16(O8)/-/-/   9

INSERT: a control
        on test equipment

e.g. turn on an oscilloscope
e.g. turn off a seven segment display

A5/O4(O6)/O1/   9
TRIM: a connector
        on an electrical component
        with a hand tool

e.g. snip off the excess wire on a resistor with a pair of wire cutters

8

The life and times of **ded**, text display editor

Richard Bornat

Harold Thimbleby

8.1   Introduction

Ded is a text display editor designed by computer scientists. The design is characterised by simplicity and adherence to user interface principles. This has led to a good design but with contentious features. This chapter highlights the conflict between principles and features, particularly in the social context in which **ded** was designed.

The primary intent of this chapter is to present the context and design decisions lying behind a particular interactive system. Neither the design itself nor its description here has been directly influenced by the book's framework; it seems more honest that way. However, the following salient points may be noted:

- Science is considered as a form of explicit, public knowledge. The public (i.e., user) accessibility of explanation of **ded** was a criterion of satisfactory design. As designers we also explained **ded** to ourselves by way of principles; thus, the role of principles was paramount, as we hope will be made clear in this chapter.
- The designers of **ded** were embedded in the environment for which **ded** was designed, though with certain concessions for some users at the periphery. Thus the acquisition representation was phenomenological.
- The development paradigm is not seriously tested by a close-coupled two-designer system. In our case, the various distinctions (analysis; generalisation; application; synthesis) were not explicitly formulated, and need not have been for a successful conclusion.
- The design of a practical system is clearly an engineering endeavour. Our design of **ded** was, however, analogous to a scientific endeavour: we used principles (qua scientific representations). We

made predictions about interaction between principles, explored them, and synthesised new principles.

Ded is a text editor that was designed at a time when computers were becoming much more accessible to certain groups of people, mostly located in university departments of computer science, who found that they were using computers for long periods at a time, perhaps for the whole of their working day, with no practical limitation on the amount of 'computer time' they consumed other than a vague obligation to be fair to other simultaneous consumers. At the same time the hardware they touched—the 'terminal' was changing as slow hardcopy printer/keyboard terminals were replaced by much faster visual display terminals (henceforth VDTs).

Ded has been very popular with the people who have used it, but many who were already used to more sophisticated editors have never been tempted to change allegiance. We do not know whether this is because **ded** is really worse for the applications those programmers have in mind or because people imprint on their first editor. There are many other plausible explanations, such as peer pressure and the so-called learning paradoxes caused by narrow horizons where users want to get their present work completed *now* in preference to investing the time to learn a new system which might improve their *future* performance (Carroll, 1987).

Certainly, some people have found **ded** good enough to go to the trouble of porting, which is no easy task. **Ded** has been ported to many machines and has formed the basis for a study in the formal specification of interactive systems (Sufrin, 1981).

8.1.1   *Why write about a 1970's system?*

Ded is years old; and user interfaces, their sophistication and our understanding of them, have moved on. Why write—or read—about it?

First, **ded** is worth talking about. It is a good system, and we want to try and explain why.

We suspect that a lot of software products are designed along the lines that if something is imaginable and possible then it is certainly permissible, and may as well be provided as another feature of the program. With **ded** we wanted to provide more than was possible (we were using small machines by present standards): we had to do less than this, and we turned to principles for guidance. Of course, we had to invent the principles too. This design strategy—analyse the task and its constraints, devise theories (we called them principles) and then synthesise a design consistent with that theory—is basically the theme of this book. Following design principles is essentially 'top-down' design; but of course, devising the principles them-

selves is 'bottom-up' (plus experience and inspiration). Design by searching for principles is a method approved by Descartes, "Each problem that
I solved became a rule which served afterwards to solve other problems"
(Œuvres, vol. VI, Discours de la Méthode). It is interesting to see how it
works out in practice.

Ded is a very simple system—which is very hard to appreciate by *reading*
about it, especially since we have emphasised design details rather than the
subjective feel of using it. **Ded** doesn't have many features, but an enormous effort went into its design. In fact, **ded** existed in almost complete,
but unpolished, form early on: we spent most of the design time (about
twelve man-years part-time, three or four full-time) improving and polish-
ing **ded** rather than adding new features. We wanted to perfect rather than
camouflage and palliate, which would indeed have been a much easier route.

Because **ded** is so simple we can talk about a significant and useful part
of its design in a short chapter. Nevertheless we have hardly done full justice
either to history (there is a lot that is not recorded here) nor have we
done full justice to **ded** itself (there are many design issues not addressed
in this chapter). As we explore design alternatives in this chapter, social
influences, principles, hardware limitations, user tasks... try and imagine
how much more difficult it would be to design a big system—say like an
airline reservation system—*really carefully*. The most interesting thing in
design is not the principles, but how they are compromised, because they
surely will be—indeed, it has been suggested that complex design cannot
be done properly but only faked (Parnas & Clements, 1986).

Like many other computer scientists we believe that programming has
nothing to do with computers: that it is an abstract business, to do with
the description of the behaviour of abstract machines. Equally it seems to
have nothing to do with social factors. But the design of software products is heavily influenced by the characteristics of computer hardware and
the ways in which they are commonly used. Thus, we are interested in
exploiting technology to its utmost. VDTs don't cost money to run, because they don't print on paper, and they work fast enough to be a plausible alternative to paper and pencil. By using up lots of computer time
you can make them do some very interesting tricks indeed. The combination of almost unlimited computer time with free printing at the terminal
made tempting overtures to a software designer. **Ded** was designed to play
the nicest tunes we could on the best hardware we could lay our hands
on.

8.2   Ded: a quick overview

Text editors are as-it-were 'stripped down' word processors, without any provision for other office procedures such as mail merging, printing, drawing diagrams, checking spelling; they are possibly one of the most widely used forms of interactive system. They are most often used in environments where the text, once edited, will be further processed by other programs (e.g., text formatters, spelling checkers, compilers etc).

The main feature of **ded** is that it is so simple it can be used straight away in almost complete ignorance of its full repertoire of commands. It is not a 'space-cadet' system. You cannot type a single key and expect to get to the end of the known universe. You *can* get there if you really want to, but you cannot get there quickly by a simple typing mistake!

The basic operations of **ded** can be presented very simply. The keyboard has keys labelled with up, down, left and right arrows and a RUBOUT (or DELETE) key. The explanation to the novice user would say that you type normally, you use the arrow keys to move up, down, left and right in the stuff you are writing and use the RUBOUT key to remove bits you don't like. When you have finished you press the ESCAPE key, type ok and press RETURN. That explanation is all you need to be able to construct letters to the bank manager, simple programs and so on. If you need to know more you can ask somebody else. The rapid feedback (for *every* key press[1] indicates to the user exactly what they are doing. In summary, you can use **ded** like a magic typewriter.

START LINE $<<<$	SCROLL UP	END LINE $>>>$	
WORD LEFT $<<$	UP $\uparrow$	WORD RIGHT $>>$	GO TO COMMAND LINE
LEFT $\leftarrow$	CENTRE	RIGHT $\rightarrow$	EXECUTE COMMAND
NEXT LINE $\hookleftarrow$	DOWN $\downarrow$		NEW COMMAND
SCROLL DOWN			

Figure 8.1: Typical keypad arrangement

On some terminals, such as those modified at Queen Mary College (by Ben Salama and Derek Coppen), the user may have further hints from the labels on other keys—for instance that the cursor may be moved in units of words, or that things other than the character under the cursor can be deleted. The 'numeric keypad' to the right of the main QWERTY keys on

---

[1]If a key does nothing visible to the user, it *really* does nothing (but it rings a bell to let the user know).

PC-style keyboards, may be arranged for **ded** as in Figure 8.1. Sometimes the keys will be labelled, but on some terminals the user has to memorise where each key is; they will be laid out fairly logically so this is not difficult. For the PC keyboard, pressing the shift key labelled ALT causes the motion keys (such as word left) to delete what they would have moved over. Thus it is possible to delete left and right in units of character, word and line from the cursor position, as well as to delete lines up and down. The central key scrolls the screen so that the cursor is positioned in the middle, to provide the user with equal amounts of context above and below the current line. This sort of keypad arrangement, combined with **ded**'s carefully-designed word and line motion behaviour (Thimbleby, 1981) greatly facilitates touch typing—none of the awful hand-repositioning and cursor-overshoot of a mouse-based system.

A user invokes **ded** by specifying a file to edit. **Ded** reads the file (or creates it if it does not already exist) and shows a display something like Figure 8.2; though the text displayed would normally be as large as the terminal screen, typically around 25 lines, rather than the 9 we show here.

```
The main feature of ded is that it is so simple
it can be used straight away in almost complete
ignorance of commands.
All a user needs to know is that what he types gets
inserted into the file, that arrow keys move the cursor around
(so that he can insert text at any point),
that the delete (or rubout) key deletes text, and
lastly how to leave the editor using the 'ok' command.
>
```

Figure 8.2: Screen layout (on a small screen!)

The screen is split into four regions. The top right region, the biggest of the four (about twenty lines deep and seventy-eight characters wide on typical terminals), displays the user's text. The cursor is shown in Figure 8.2 at the top left hand corner: the arrow keys can be used to move it anywhere within the region. The bottom line is a command line, where the user can type special commands to **ded** that are not intended to form part of the text.[2] For example, 'ok'—the command to leave **ded**—would be typed on the command line. The '>' symbol is fixed, and helps to make the command line appear separate from the rest of the display. The left hand column is for more advanced use: it is used for naming regions of text. A special command names a line or several lines of text; other commands can

---

[2]The command line is on the bottom rather than the top for reasons discussed in Granada and Teitelbaum (1982)—though we had already decided before reading the paper.

be used to do their work within such a region of text rather than on the entire file.

The lines between regions in Figure 8.2 don't really appear on the screen. Figure 8.3 shows what would actually be shown after the user has named a few lines 'a' and is about to execute the 'ok' command to leave **ded**. The

```
    The main feature of ded is that it is so simple
    it can be used straight away in almost complete
    ignorance of commands.
a   All a user needs to know is that what he types gets
a   inserted into the file, that arrow keys move the cursor around
a>  (so that he can insert text at any point),
a   that the delete (or rubout) key deletes text, and
a   lastly how to leave the editor using the 'ok' command.
>   ok_
```

Figure 8.3: Screen layout as really seen

bold '>' (darkened in this book, but bright on white-on-black terminals) mid way up the screen indicates the line the cursor used to be on, before the user moved it to the command line at the bottom.

You might mistype 'ok' (as 'ko' perhaps). The command that you type is visible on the screen so that if the editor didn't recognise it you can at least read it to see what might be wrong. It is natural to consider that it might be changed: indeed, it can be treated just like the main text and by using the same keys. You can insert and delete characters within a command and then ask for it to be carried out.

Complete novices can use **ded** successfully after a few minutes exper-imentation. They have much more trouble with the QWERTY keyboard, with the idea of files and of storing texts, with the notion of constructing UNIX commands in sequence, than they do with constructing and mod-ifying texts with **ded**. The idea of using programs—like **ded** and text formatters—to process text held in files is difficult to grasp at first. The fact that UNIX commands don't mean anything to **ded** and **ded** commands don't mean anything to UNIX is confusing too, and the conventions of most formatters are incomprehensible to anybody in the limit. **Ded** is simple because it concentrates on a part of the documentation processing job, because it can only be approached by somebody who has understood the basic mystery of file storage and is prepared to be introduced to a corner of the impenetrable mysteries of formatting. In part we are hiding behind the complexity of other parts of the UNIX system: we could make **ded** as simple as we wished, so long as some bit of UNIX can be relied upon to do the rest of the job.

There is no 'help text' in **ded**, although we were often urged to provide some. We were and are quite happy with the fact that naïve users of **ded** use only a portion of its facilities and have no idea that the rest of it exists. They find out more about it from talking to other users who know more— **ded** was designed for a timesharing environment where you are never far away from somebody else who is using the same machinery. After a few years we wrote a full manual—it ran to about seventy pages and the only thing we remember about it is that some wag said that pencil, paper and rubber don't need SEVENTY pages to explain.[3]

8.3   Historical background

Historically, text editors were heavily oriented to the medium carrying the text. Thus, in the 1960s and early 1970s paper tape was a quite popular medium for storing text: users could edit paper tape by taking scissors, poking holes and using masking tape to obscure unwanted holes. The user clearly had to be *au fait* with the binary character code, but otherwise the operation of the 'editor' (i.e., splicer and punch) was 'direct manipulation'. The tape itself did not record the history of the editing: all that could be seen from the tape was the *new* text (and some bits of masking tape). Searching for the right place to fiddle was greatly aided by running the tape row-by-row through a reader/printer/punch terminal so that you could read the corresponding text.

It soon became popular to use interactive computer terminals. Now the computer, rather than the user, could perform the searches to find the places in the text where changes were wanted. Edits would be specified by instructions: for example, to delete a line of text or to replace a line with some new text. Hardcopy terminals, such as electronic typewriters, make a record of each instruction. Now the record of an edit is the *history* of the changes: the sequence of instructions that together dictated the edit. Here is an interesting change brought about by the technology: on-line terminals certainly support a more flexible and easier-to-use way of text editing, but the representation of the editing is now more complex. For many users the history of the editing became distracting—a price to pay for the extra flexibility. Figure 8.4, a typical editing session in this style, cannot be understood without knowing the detailed history of the user's interaction; indeed some of the text might even be the result of an earlier print command (i.e., the apparent interaction might be text in the user's file) and then you would be thoroughly confused.

---

[3]Writing the manual helped uncover a few minor bugs, so it was not a totally wasted effort. On the other hand, if we had not written the manual, perhaps nobody would have even thought of trying the things that did not work properly or even known what they should have done!

The history could exhibit lines of text in any order on the paper: the fact that one line of text preceded another would not mean that it *textually* preceded the other, merely that the user had *examined* it earlier. It was easy for the pressurised user to confuse the two and make mistakes. There is a trade-off to be made, between the simple-but-tedious text editing of the paper tape era and the complex-but-fast text editing of the on-line terminal era.

When the visual display terminal became widely available in the late 1970s it became possible to reappraise the design of text editors. Now it was possible to display a page of text on the screen and to show how it changed under each instruction to edit it. Just like the old paper tape editor, there was no need to actually show the sequence of instructions used for making the changes.

Such text editors should be easier to use (assuming that the user is not especially interested in the history of their session)—so long as users can reliably construct the commands for editing the text without knowing the history of previous edits. Since the editor does not display the history, there should be no reason for the user to have to remember it.

Conventionally, a 'history-free' interactive system is called *declarative*, to distinguish it from a *procedural* (or *imperative*) interactive system. Very often it turns out to be difficult to provide a purely declarative system—the compromises are often called *modes*, being information the system keeps to itself. There is some debate—that will be touched on in this chapter—over the merits of modes: some modes for certain purposes appear to be necessary and useful; others appear to be superfluous and damaging for the user. Modes mean that the user has to do or say less, but run the risk

```
/changes/
and show how it changes under each instruction to edit it.
-
Now it was possible to display a page of text on the screen
+
and show how it changes under each instruction to edit it.
s/changes/changed/p
and show how it changed under each instruction to edit it.
/othe /
Unlike many othe Unix tools it is generally
s//other/
?on-line computer?
It soon became popular to use on-line computer terminals.
c
It soon became popular to use interactive
computer terminals.
```

Figure 8.4: Transcript of an editing session
Notice that it is not obvious what text is typed by the user

that the mode information hidden in the computer is not what the user thinks it is. The fewer modes, the more declarative a system and (as is now widely argued in support of 'fifth generation' computing projects, such as Prolog) the easier systems should become to do powerful things.

Ded is a declarative low-mode editor: quite a novel idea for its time. It was so novel a concept, we did not know what it was called when we did it; we called it *picture editing*, an idea that we discuss more fully below in section 8.6.5. The declarative nature, and our deliberate adherence to it, makes **ded** radically simpler than every other editor known to us.

8.4   Design versus evolution

The most important thing about the design of **ded** is that it never happened, in the sense that there was never a time when a group of people—computer scientists or ergonomists or anybody else—sat down together and thought about what the new editor product would be like. Perhaps it would be truer to say that there was no *one* time when we sat down together and that we didn't sit down together *before the editor existed*. There were lots of times when people sat down together. There was a lot of design activity, but it didn't precede production, it suffused it. It was rather like what is nowadays called 'prototyping' except that in our case there was always a fully working version of the product to be discussed.

We like to believe that **ded** wasn't designed, it evolved. Richard wrote the first version because he needed it for his work, colleagues begged to be allowed to use it as well, and the bandwagon was rolling. We worked out design principles during the evolution of the design: **ded** was rewritten at least three times to fit new ideas of what those design principles should be, and at least one of those rewritings was thrown away when we reverted to previous versions of our principles. Our evolutionary process was a sort-of genetic engineering rather than natural selection.

Our notion of the target population of users changed over time. From a program written for one person to use, we eventually believed we had developed one which could be useful to the secretarial staff in our departmental office, who spent too much time retyping documents which could easily be word processed. Nowadays they use **ded** all the time. Later still we believed it would be useful to novice undergraduate computer scientists, who have problems understanding just about everything they come into contact with.[4] We were much less responsive to the needs, or at least the expressed

[4]We also had some idea that the program would be too computationally demanding to be used by lots of students sharing one machine. But it seems that the students actually spent less time compiling and correcting incorrect programs: overall they were better off.

demands, of fellow programmers: they could write their own editor if they didn't like the way we did things.

Social factors helped to make the design process evolutionary. The Computer Systems Laboratory at QMC tried to be a non-hierarchical institution, at least so far as academics and research students were concerned. Nobody felt empowered to say what other people should be doing. Sometimes they felt they should tell other people what they should *not* do, for example because they felt it was wasteful of lab resources, but Richard's position as a senior academic meant that he could get away with most things. So we could design **ded** without asking anybody's permission beforehand and we could make it do just what we wanted without taking notice of anybody else's feelings or opinions.

Computer programmers don't keep their opinions to themselves very much and we were often informed of **ded**'s supposed shortcomings and ways in which they might be overcome—sometimes we feel that we talked about little else, though Richard made himself responsible for design decisions. That made **ded** a very 'principled' program. If he thought a suggestion to improve it didn't fit with the current set of design principles, he didn't make that improvement unless and until he could be persuaded. He was often persuaded, but sometimes obdurate.

Ded was very much a product of us and our environment. We don't know what we would have done somewhere else, or what we might have done if we had had a different idea of the target users. Different approaches are possible under different circumstances: the design and motivation behind **ded** (partly covered in Thimbleby, 1981, 1982 & 1983) can be contrasted with that of **sam** (Pike, 1987) and EMACS (Stallman, 1984). Meyrowitz and van Dam (1982) provide a general survey of text editing.

8.5   Early developments

Richard was in the middle of writing a book when he came to QMC in 1977. This was the major impetus for the design of **ded**. The rest of this section, section 8.5, describes Richard's background; it has been written in the first person.

My design ideas have come largely from experience of and dissatisfaction with other people's software products. In the case of **ded** *my dissatisfaction was with the way that contemporary software appeared not to exploit the capabilities of the new VDT technology and to waste the free computer time which I was beginning to enjoy. I had already accumulated about a hundred pages worth of text in machine-readable form. I wanted to be able to review this text, to modify it and to add new text, to reorganise its structure of sections, chapters and paragraphs, to write whole new chapters. I wanted* to use what is nowadays called a word processor and since nothing like that existed I had to invent my own.

I had constructed my hundred pages in the first instance by using an imperative line-based editor, though in the later stages I had used a declarative picture editor which displayed twenty lines or so of text on the screen and allowed simple overstriking and insertions. All that had been done using what now seems ridiculously inadequate hardware, which could print between ten and thirty characters per second. The hardware at QMC could handle up to a thousand characters a second. I could see that it would be possible to compose text on screen, to type paragraphs and to read them in their context without having to specify what the context was and for the first time to move upwards in the text, to show what was off the top edge of the screen.

In inventing my machine for writing the book I drew upon experience with all sorts of earlier text editors which I had used. Major influences were the line-based editor **SOS**, which I had used on the DEC PDP-10, my experiences in using an editor called **SYRUP** I had written for my own use on the PDP-10 and **em**, which was a version of the UNIX text editor **ed** devised by George Coulouris running on the PDP-11. Apart from **SYRUP**, all of these were imperative history-based editors.

Like many other people at the time, I was fascinated by the possibility of producing 'typeset' text. UNIX had and still has programs called **nroff** and **troff** which do this job. They work with text which is interspersed with formatting commands: for example the previous paragraph was produced from the text shown in Figure 8.5.

```
.PP
\f2In inventing my machine for writing the book I drew upon
experience with all sorts of earlier text editors
which I had used.
Major influences were the line-based editor \f4SOS\fP, which
I had used on the DEC PDP-10, my experiences using an editor
called \f4SYRUP\fP I had written for my own
use on the PDP-10 and \f4em\fP, which was a version of the
\e*U text editor \f4ed\fP running on the PDP-11 devised by
George Coulouris. ...
```

Figure 8.5: **troff** typesetting commands

The most important thing about text to be given to a typesetter is that the original layout hardly matters. The typesetting program works out where words are to go on the page, places headings, decides on margins and so on: the result is quite different from what you type. When original textual layout does matter it is in order to place particular characters at particular positions on the page, to produce a particular character-picture *in the formatted text, and character-pictures can quite easily be built up character by character. That meant that I could use a 'no-frills' text editor to construct my book, an editor without facilities to arrange and format words on a page.*

8.5.1   Idea: Only one mode[5]

I wanted to be able to compose English text on screen. I wanted to survey the text as I typed it, so I wanted it at every instant to show me exactly what I had typed. I didn't know what to do about insertions and overstrikes—that is, whether a key depression should signal the editor to replace the character at the editing position (overstriking) or to insert a new character before the one at the editing position (insertion).

At the time, because of limitations in VDT hardware and shortages in computer time, many declarative (history-free) editors worked normally as overstrike editors, but could be commanded to make an insertion: then they inserted a wide space in the text into which the insertion was typed by overstriking, closing up the gap when the insertion was completed. I didn't like that idea because it meant that you couldn't easily read a sentence you were modifying: the big space made it hard to read and quite often part of the sentence fell off the right-hand edge of the screen. I'd experimented, in the editor **SYRUP**, *with a more computationally expensive insertion technique which put in a character every time you pressed a key, and in particular rubbed out the last character you had typed whenever you pressed the* DELETE *key. But I didn't know what to do about deletion when the editor wasn't in insertion mode.*

Here a happy accident intervened. In a conversation with Jon Rowson, one of my colleagues in the lab, I was introduced to the idea of a 'modeless' editor, one which always worked using insert/delete and never allowed overstriking. It just seemed a good idea, and I adopted it immediately.

In order that I could type rapidly I made the editor put in an automatic line-splitting command whenever the text got close to the right-hand margin of the screen, but I made no further concessions to layout.

8.5.2   Idea: Search for anything

The hundred pages of book text that existed in 1977 were in **runoff** *format:* **runoff** *was the remote ancestor of* **nroff**, **troff**, TEX *and countless other typesetting programs. QMC's* UNIX *machine had* **nroff**, *and* **nroff**

---

[5]Pedants may notice that we use the terms 'one mode' and 'modeless' apparently interchangeably. Technically a system must have at least one mode, but we can understand a system appearing to be modeless if its user can cease to be aware (and *not need* to be aware) of the one mode it is in.

would enable me to typeset pages which were much more readable than **runoff**'s *attempts. Both programs use typesetting commands which appear on single lines within the text to be set, but their command sets are different. In one very important and influential case—footnoting—it was necessary to change more than one command at a time.* **Runoff**'s *footnoting commands were as shown in Figure 8.6* **Nroff** *recognised more compact commands and allowed automatic footnote numbering, so I wanted to alter that text to the form shown in Figure 8.7.*

```
information is prepared about the arguments[8.1]
.FN9
.B1 .F .NJ
-----------------------------------------------------------------
[8.1] I have used 'argument' to describe information passed in
the procedure call and 'parameter' when used by the called
procedure.
.END FOOTNOTE
of the call;
```

Figure 8.6: Footnoting in **runoff**

```
information is prepared about the arguments\c
.fn
I have used 'argument' to describe information passed in
the procedure call and 'parameter' when used by the called
procedure.
.ef
of the call;
```

Figure 8.7: Footnoting in **nroff**

None of the text editors available at QMC could recognise patterns which extended outside a single line, let alone do substitutions which altered more than one line of the text. I wanted to be able to change my text systematically and I believed that it would be easier to write a program to do it for me rather than to go through the text laboriously, and probably unreliably, changing it piecemeal.

So I decided that I wanted to be able to search for and replace textual patterns which extended over a single line. I found that it was very difficult to type the commands which recognised something as complicated as the footnote pattern illustrated above. In its first incomplete form, whose notation took some time to evolve, this command would be as shown in Figure 8.8 below.

I don't think it's necessary to give an explanation of this notation—just feel the complexity!—but Appendix 3 gives some explanation for the curious.

```
x/['[-]']'*]'^'.'*'^'.'*'^----'.'*'^['[-]']'*]/\c'$.fn'$
```

Figure 8.8: A ghastly search & replace

8.5.3   Idea: Make everything visible and editable

Many editors do not make the command line visible; but if **ded** allows commands as complex as in Figure 8.8 then I needed to be able to see what I was doing! Once I had constructed a command which was recognisable to **ded**, it was probable that I hadn't constructed the right one. Usually I'd got the search pattern wrong, so that it didn't find anything in the text to replace. So I wanted to be able to edit the command line, using the same editing features as for editing other text.

8.5.4   Idea: Make changes interactively

A more dangerous possibility was that my command described a replacement of some sort, but not the one I intended. That made me design an interactive replacement command which would show the effect of each replacement, in its context, for my confirmation or rejection. I borrowed this idea from the line-based editor **SOS** which I had used on the PDP-10.

8.6   Design principles

Because **ded** evolved, the design principles which we believed at the end weren't the ones we started with. There may be important features of the editor which are only explainable in terms of principles which we have forgotten. So far as we can tell from the notes we made at the time, we did have a number of principles in mind. Some of them may have been little more than platitudes, but we believed we followed them at the time—and we still believe in most of them now. The ones we believe in are described below.

8.6.1   Always show something is happening

When editing text something happens every time you press a key: a mark appears or disappears from the screen, the editing position cursor moves or in the extreme the terminal bleeps. Sometimes a command, particularly a search in a large file, can take a long time to carry out. It was very important to show that **ded** was working away in the background, even when the screen wasn't changing. We made it display a white blob mark at the end of the command line after hitting RETURN. Very simple, but quite important, not least because it dissuaded people from pressing the RETURN key again and again in the belief that **ded** hadn't noticed them the first time, which could produce nasty effects when the command finally finished.

8.6.2   *Show the context, but keep the screen still*

Because we wanted to compose and modify text on-screen we wanted always to see the context within which we were working. We wanted always to be able to read the text around an insertion, because what we were typing would typically be some sort of extended explanation of something else that was already in place. The problem is that 'context' is maintained by **ded**, but exactly what the context is depends on what the user is wanting to do—and **ded** knows nothing about that.

Showing as much context, text above and below the cursor, as possible seems a sensible strategy. It would seem to mean redrawing the screen image so that the insertion point was on the middle line of the screen, and making sure thereafter that it stayed there as the user moved the cursor around. In practice this had some disadvantages. If the screen image jumped about whenever one started an insertion it didn't feel like typing on a piece of paper. Also, it was by no means true that the relevant context for the user was equally above and below the insertion point: it might be mainly the text above or mainly the text below that was important. So we rejected the notion that insertion should trigger a change of context.

If it is important to display as much context as possible from moment to moment, then we should at least arrange for the user to be able to take full advantage of what context he can see. That is, we should keep each context as long as possible, only changing it when really necessary. How to keep context was more difficult to decide. It seemed that to move down the screen as the text grows felt more like a typewriter, more 'natural'. We rationalised this to say that users would prefer modifications, whether insertion or deletion, to leave as much as possible of the screen unaltered. With hindsight we may have been influenced by the limitations of our hardware, so it may have been a concern to update the screen as rapidly as possible which was paramount. At any rate the principle 'keep the picture as still as possible' became a design principle and is even preserved as a comment in the text of the program—this is Hansen's principle of *display inertia* (Hansen, 1971).

Trying to show as much context above and below as possible obviously conflicts with trying to keep the display as still as possible. In the limit, using the techniques employed in **ded**, you reach the top or the bottom of the screen. We decided that close to the screen margins it was more important to preserve the typing context than to make editing instantaneous. We kept at least two lines above and below the editing position visible—hang the response delays and the movement in the picture. Because an attempt to move the cursor into the top or bottom margins causes the screen to scroll and by doing this **ded** generally centres the cursor, the next scroll (and the delay it causes redrawing the screen) will be postponed as long as possible—until the user has managed to reach a screen margin by moving from the middle.

8.6.3   Fast interaction

It was important that typing new text to the editor should be as fast as typing on a typewriter, or else it would be better to be working with pencil, paper, scissors and sticky tape. The delays involved in processing commands from the keyboard and sending commands to the screen were sometimes quite severe. We found that when our time-shared computer was overloaded the delays were so extended that it was impossible to work unless you could touch-type with your eyes closed. The only rational policy was to make the delay as close to zero as possible and then hope that it was small enough.

We designed an editor which would do all the things that we thought it should and made it go as fast as possible on the hardware we actually had. When we got more capable hardware, we altered **ded**'s internal workings so that interaction became faster, but we didn't change its capabilities at all.

8.6.4   Ignore deficiencies of the terminal hardware

When **ded** was being written most of the computing world had standardised on the ASCII character set of ninety-six printable and thirty-two control characters, but it hadn't standardised on how VDTs should work. (Now it has, more or less, but oh dear, that's another story!) The VDT on which **ded** first operated could recognise commands to overstrike a character, to move down a line and to move the editing position (the 'cursor') to any position on the screen. The 'move down' command worked like a 'line feed' when the editing position was at the bottom of the screen and moved the whole picture up one line. The VDT couldn't recognise commands to move up a line, to insert a blank space or a blank line or to pan the picture left or right. Despite these deficiencies it was quite easy to write a program which would produce any desired change of a displayed piece of text.

One design principle, first articulated to us by Bruce Anderson, was that **ded** shouldn't be designed *not* to do something just because it was difficult or slow to produce that change on our particular VDT hardware. It was clearly desirable to be able to insert a character or split a line anywhere within the displayed text: it would be intolerable if it could only be done on the top or the bottom line of the picture, for example, even though those might be the only places where the VDT hardware could do it fast enough to be apparently instantaneous.

8.6.5   *Edit a picture of the text*

A central decision in **ded**'s design was to be declarative—the user edits a picture of his text. "Editing a picture of your text" is how we would describe the idea to a user (and described it to ourselves at the time); "declarative" is how we would now describe the idea pompously. The dual form of the idea constitutes a *guep* (Thimbleby, 1984).

If the user edits a picture, he can only edit what can be pictured! If the text contains formatting characters like newlines, tabs, blanks and so on—as it inevitably will—characters where there is no one-to-one correspondence between the text itself and the picture it describes, **ded** has to choose a pictorial convention that can be edited directly as picture. **Ded** provides no way whatsoever of editing the text itself, that is the ASCII text, directly. For more background to this decision see Appendix 2.

An immediate and contentious consequence is that **ded** discards blanks at the end of lines of text, and tab characters are treated as multiple spaces. The newline character has no significance for picture editing; the user can position the terminal cursor anywhere on the picture without regard for the corresponding notional cursor position in the text it is representing. Thus the cursor may be placed 'beyond' the end of a line without special commands. A pleasant consequence, unfamiliar to users of some other editors, is if the '↑' key is pressed repeatedly the cursor moves up vertically (rather than zig-zagging along the end of lines).

```
A pleasant consequence, unfamiliar
to users of               _
some other editors, is if the '↑' key is
pressed repeatedly the cursor moves
up vertically             _
(rather than zig-zagging along the end of lines).
```

Figure 8.9: Possible cursor positions moving upwards in **ded** (starting from the 'i' in 'zagging')

```
A pleasant consequence, unfamiliar_
to users of_
some other editors, is if the '↑' key is_
pressed repeatedly the cursor moves_
up vertically_
(rather than zig-zagging along the end of lines).
```

Figure 8.10: Possible cursor positions moving upwards in a silly editor (again starting from the 'i' in 'zagging'). Once the cursor 'attaches' to the end-of-line it stays there!

Notice the non-declarative (history sensitive) nature of the design represented in Figure 8.10. Suppose at some stage during an interaction the cursor is at the end of the second line (i.e., just after 'of'). When the user moves the cursor up, it will either move to 'c' on the line above or to just after 'r': depending on the user's earlier actions and whether there are invisible spaces to the right of the 'f' on the second line. A popular compromise is a design half way between Figures 8.9 and 8.10: and is worse than either. The cursor would move as in Figure 8.9, but as soon as the user types a character, it has to go in the text, so the cursor jumps to the corresponding position as in Figure 8.10.

Further advantages can be deduced from Appendix 2, and include the ability to edit the full ASCII character set—a principle that has section 8.6.6, next, to itself.

8.6.6   *Full ASCII character set*

Programmers like to be able to type control characters (the ASCII code allows for one hundred and twenty-eight characters, but only ninety-six are printable text). Since it is extremely inconvenient to be unable to edit and correct a *corrupt* text file (for that is what we shall call a file with normally-unprintable character codes in it), **ded** should provide some means to edit non-printing characters. Of course, 'corrupt' files arise most frequently when programmers are at work! **Ded** uses a convention that a character following '◇' is converted to a control character of the same name. Thus '◇a' (or equivalently '◇A') represents the unprintable control-A.

Since **ded** is a picture editor, the user can treat '◇A' as two separate characters. For example, he could move the cursor between '◇' and 'A' and insert a 'G': this would create control-G followed by A. Most usefully, the user can search for *any* control character by searching for '◇' directly—it is simply treated as any other character on the screen.

8.6.7   *Fit in the memory of a PDP-11*

Ded had to run on the hardware we could afford, because we needed to use it. The PDP-11 was a very small machine and, in the early days, programs and data had to fit into 64K bytes. This was always a tiny space. It meant that we had 'hardware support' for any refusal to add new facilities to **ded** because there was never room in the memory space available. The only way to add new stuff was to find out ways to generalise what was already there, to make the new addition operationally and conceptually consistent with what was already there.

Because **ded** evolved it was always rather a messy program and there were always ways in which it could be rewritten as a smaller program to do the things that it could already do. So there was often room for an improvement if we *really* wanted to make it, but the constraint of space was always a potent one.

Note that the PDP-11's limited address space did not favour particular features over any others. If we had had a different sort of computer it might have had curious hardware that restricted certain aspects of **ded**'s design more than others. In some ways we were fortunate to have an impartial restriction imposed on us: without the restrictions it would have been far easier to extend **ded**, adding features, rather than get what we had right. So we were forced into 'survival of the fittest'.

8.6.8   *Simple operations should be simple, and the complex possible*

We've seen that **ded**'s operations are simple, but we bowed to programmer's pressure by providing global replacements, which can occur without visual confirmation of effects.

Global replacements are repeated replacements made by the editor throughout a document (i.e., globally). For example, the user may wish to correct the spelling of 'commitment' everywhere he spelt it 'comittment'. If we follow our principles, the user should be able to see each change. We might also want the user to confirm each change the editor proposes to make. However, if a document is very long, watching (and possibly confirming) each change can be very tedious. Programmers were particularly insistent on the need to avoid tedium even at the expense of security and as a compromise, **ded** has two forms of global replacement— one makes changes one-at-a-time, querying the user at each step; the other performs all changes 'at once' unconditionally.

8.6.9   *Be safe to use*

We do not believe, still less require, users to be perfect typists. **Ded** follows a 'shaky fingers' model: typing mistakes are easily seen and corrected. Some typing mistakes or slips (such as trying to move left when the cursor is already as far left as its going to go) cause the terminal bell to ring. The bell also helps copy typists, interviewers... people who may not want to pay full attention to the screen. There are good psychological reasons for using sound to notify users of simple errors, though there are equally good sociological reasons to prefer silence (do you want the whole typing pool to know your error rate?), so **ded** provides a command to silence the bell if necessary.

To complete an edit, type 'ok'. To abandon an edit, type 'q'; but if the file has been changed it is sensible to require some sort of confirmation before abandoning possibly hours of editing. Many editors ask the user to confirm by typing 'y' (for 'yes'), or to type the q command again—but it is all too easy to slip into a habit of typing 'qy'! Instead **ded** requires the user to type the long command '`reallyquit`', in the hope that one is less likely to type it by accident.

We have seen that some features in **ded** permit a happy co-existence of expert- and naïve-user oriented features. Many editors provide 'levels' or other features for customising to specific user types; levels can be provided very easily by a programmable user interface, such as that provided by the editor EMACS (Stallman, 1984). It is clear that this is a very useful feature for programmers, most especially in research and development environments. But it is also clear that customisation, even when not as radical as might be achieved by programming, can change a user interface (of course, this is what the programming is for!) so much that a user cannot rely on his colleagues for help—for he uses a *different* editor. When does a customised editor become a different editor? Is customisation a designer's let-out for an incomplete design?[6]

The user interface of **ded** is fixed and all features are available to all users at all times. Interestingly, we have not yet found a programmable editor that could emulate **ded**: most serious problems arise with the notion of picture editing and cursor motion (EMACS, for instance, insists on editing ASCII text). Some operations, such as word motion, become too inefficient even if they can be specified.

8.6.10   *Undo anything*

We both felt that **ded** should have an undo command. Although Harold left QMC in 1981 and we implemented undo independently, our ideas were remarkably similar: sufficiently so for the following paragraphs to correctly describe *both* approaches.

It is remarkably easy to delete a line by mistake, and then you cannot see what it was, let alone remember it to retype it. Undo commands allow the user to recover from such disastrous mistakes. Our philosophy was that it was better to be able to undo anything, even if the user had to wait a little while, than (as most other editors work) undo only certain things on certain days (but perhaps faster than our more general scheme). Some editors can only undo the last thing you did (if it is the sort of thing that is undoable!); often you cannot undo more than one thing, because the editor is arranged so that an attempt to undo two things just undoes the first undo.

A constraint on the undo command was that it must be simple (we don't want the user making mistakes with the undo command that in turn have

---

[6]To appease those people who like customisation—it *is* a let-out, but then designs are *never* complete...

to be undone by another error-prone meta-undo command), and that it should fit in with the declarative low-mode nature of **ded**. The mode problem with undo commands is quite serious: the undo command can mean *anything*—the reverse of whatever has to be undone. The solution was to make undoing a mini-dialogue that displayed something of what commands would be undone: thus, at least, the user can see what would happen.

An important property of the undo command is that if you want to undo a lot of work, it will take you an equally long time (because you can only undo one keystroke at a time). This helps ensure that you do not make catastrophic undos which you then need to undo...

8.7   Problems with the design

There are all sorts of ways in which 's design might be improved. Indeed it might still be under improvement, still evolving, if we hadn't simply lost interest and moved on to do other things. There are certainly some ways in which its design ought to have been better.

Problems in the design can be classed as:

- *weaknesses in implementing the principles* e.g., the search notation
- *difficulties in implementing the principles* e.g., editing tables
- *unfortunate consequences of adhering to the principles* e.g., editing tables
- *other compromises made under pressure* e.g., non-interactive substitution

We shall explore these problems below.

A major disaster area is the notation for searching in the text (see Figure 8.8 or Appendix 3). Novice users don't master it beyond simple searches for words, and its adherence to *exact* character matches makes it not so useful for word processing. **Ded**'s simple search commands are simple enough, but the complicated search patterns are horrid. It *is* possible for the expert user to do almost anything, but it is very hard to work out how to do it. A regular expression grammar isn't a good programming notation and in some respects the design was far too complicated to be good. Only the two of us completely understood it and that was a long time ago.[7]

Ded does have some modes. The simplest example is that the RETURN key, pressed when editing text, breaks the current line but when editing the command line, signals the execution of the command. For example, to search for '**xyz**' you press ESCAPE, **/**, **x**, **y**, **z** and then RETURN. For a time

---

[7]At least one of the people who read a draft of this chapter claimed to understand it as well. When we have enough candidates we will set an examination.

we resisted this double use of RETURN but it is so ingrained in the mind of the UNIX user that to execute a command you type it and press RETURN that in the end we had to capitulate.

Another capitulation to public pressure is the non-interactive substitution. The normal **ded** substitution ('**x**' for 'eXchange') shows each substitution as it occurs, asking for confirmation or rejection. The abnormal command ('**s**' for 'Substitute') does them all regardless. The effect is often devastating, but that serves them right for asking for such a nasty facility.

8.7.1   *Unfortunate consequences of the picture approach*

Ded edits a two-dimensional picture of a text in which a line has position vertically and a character or a word has a position within a line. In many ways it is convenient to view a text one-dimensionally, as a sequence of characters, words or sentences. An example of **ded**'s two-dimensional fixation, taken over from the line-based editors which it supplanted, is that it is only possible to command it to mark, move or copy blocks of lines, not characters, words or sequences of words.

Long lines are messy, so it is important that they do not occur through normal editing or text entry using **ded**. To this end, **ded** provides a 'word-wrap' feature (which is a standard feature nowadays for any word processor). Simply, **ded** inserts 'start-a-newline' instructions *between words* in the stream of the user's typing whenever it gets too close to the bell margin. The effect is that the user may continually copy type, and **ded** will fill text lines well enough. In fact, *good* wraparound is a difficult feature to specify: our 'good enough' wraparound strategy tends to result in occasional uneven line lengths.

When a session with **ded** is completed, the edited text has to be reconstructed from the text picture and this cannot guarantee restoration of the original byte sequence. If the user inserted a tab, **ded** would have inserted spaces into the picture. The user may then edit these spaces or move the cursor across or around them. When the user has finished editing, **ded** assumes the spaces really are spaces and not tabs. **Ded** is *idempotent*: if a file is once edited by **ded**, **ded** changes it into a picture form (i.e., by removing invisible text such as spaces at the right hand ends of lines, removing blanks lines at the end of the file, replacing tabs by spaces) and it will then stay in that form. Furthermore, a user cannot construct a non-pictorial file by using **ded**. Thus the ASCII/picture transformations made by **ded** should be of no concern to the user (unless possibly he is worried about data compression—**ded**'s strategy penalises programmers who use lots of tabs).

Ded is awkward to use to construct tables. It is easy to construct a preliminary picture in which various parts line up in tabular form,[8] for example as in Figure 8.11

```
        Table of contents

Footnoting              44
Grumbing                46
Headaches               50
```

Figure 8.11: Tabular layout

But when changing the spelling of 'Grumbing' to 'Grumbling', the page reference on that line is pushed one place to the right. That is a trivial example of a serious difficulty which could not be solved within our set of design principles—and it gets worse when you want to use symbols like π in your text. Again, the problems can be pushed over into other software tools, in this case to **tbl**, a table formatting system that can do this and much more besides that would never be attempted by an editor.

8.7.2   Scrolling

The terminal screen **ded** uses is of course far smaller than most texts that users will want to edit: it is not as high, nor as wide. The conventional solution to simulate indefinite height is to *scroll* the image vertically on the screen. This gives the appearance of a window onto a 'scroll' of text extending above and below the physical screen. The same technique could be used to scroll sideways (like a real papyrus scroll—so the rolls don't get in the scribe's way), but it turns out that few terminal manufacturers felt it an important enough feature to provide. It was simply too computationally expensive for **ded** to support left-right scrolling: it is very slow to use, the text picture 'falls apart' during the scroll operation and it would have been tedious to program. Thus **ded** scrolls vertically, but uses a different technique to handle arbitrarily wide text.

Vertical scrolling is an interesting problem that has been widely studied. The user can move the cursor *within the screen* by the '↑' and '↓' instructions. What should they do at the edges? **Ded** takes them to have the 'same' meaning relative to the text: that is, for '↑' the screen will scroll downwards, so revealing the previously hidden text *above* the cursor. The cursor will then move up one line as required.

Wide text has to be fitted within the width limits of the screen. Since **ded** cannot assume to know what the text means (**ded** does not aspire to be

---

[8]We refer to an earlier age, when computers printed with fixed-width characters. Nowadays when printers use proportionally-spaced fonts things are much more difficult.

a word processor), it must not make an arbitrary 'line break', particularly one that cannot be seen. We can either have the text indicated as 'off the screen' (perhaps by displaying a '⇒' symbol at the right hand end of a truncated line), or we can put the continuation of the line immediately beneath:

```
...t indicated as 'off the screen' (perh⇒
...t the right hand end of a truncated
...on of the line immediately beneath:
```

or

```
...t indicated as 'off the screen' (perh◇
aps by displaying
...t the right hand end of a truncated
...on of the line immediately beneath:
```

Figure 8.12: Alternative ways to handle long lines

There are disadvantages to either solution. In the first case, if the user scrolls the screen to the right most of the contextual information—probably contained in short lines—will be lost off on the left. In the second case, at least the user can see the full context, but moving the cursor vertically is problematic: we either have to compromise the picture idea, or have the cursor jump a couple of lines at a time. **Ded** uses the fully visible, wrapped, approach but we are not sure that we ever thought it out properly. Perhaps we wanted to discourage long lines? William Newman's remark at the time that panning editors are confusing to use, because you lose context when panning along a long line as the rest of the text slips off the left of the screen was comforting. We never experimented with panning in **ded** until much later (when it was too late to do anything about it); however our experience then, and with other panning editors since, supports William Newman's opinion.

An interesting feature of the wrapped approach is that the user can use the *same* notation himself to join lines. If **ded** used the ⇒ notation, typing '⇒' (even if the user could type it) would have no effect because there would not be any text *already* to its right (and therefore **ded** should not display it). But with the ◇ notation, the user *can* type '◇' at the end of *any* line in order to join it to the following line, whatever is on either line. This is a simple example of 'equal opportunity' (Runciman & Thimbleby, 1987).[9]

[9]Not everything is equal opportunity. The 'file' command tells the user what the current file is called. **Ded** responds by displaying '>file *name*' on the command line. Unlike many other commands, the user cannot edit what he can see—the name, to

8.7.3 *Integration with other software*

A text editor is a cut-down word processor: but we still *want* all those useful features. A word processor comes with all the features integrated into the same system: this means that the users (and designers!) are faced with a larger, harder-to-understand user interface. On the other hand, a separate text editor such as **ded** has to comply—or rather its users have to comply—with the arbitrary and generally unrelated conventions of those programs providing the extra features.

It is generally impractical to run **ded**, or indeed any display editor, by other programs. One would like to integrate the editor of one's choice with all interactive systems (such as mail systems, spreadsheets etc.) that involve text input or editing. Unfortunately display editing commands generally have incremental effects on the editor state which means that programs either have to maintain an equivalent state or parse the display output of **ded**. Neither is practical, and programs that use **ded** are reduced to simply invoking it and trying to recover from any errors later.

The picture issue is also a serious impediment to integration because **ded** may be used to prepare data for other programs which want to distinguish between (say) tabs and spaces. We can escape criticism by claiming that **ded** is a *text* editor, not a *data* editor—but programmers might claim that if **ded** is so good, why can't they use it for all editing? Which sort of consistency is more important: simplicity or universality? We felt that a good user interface was more important than superficial consistency with existing programs. At QMC we had access to the source of all programs and we were able to make changes in many cases. Systems programmers elsewhere may be less fortunate or less motivated.

8.8   Conclusions

What can be learnt from studying the design activity which led to the production of **ded**?

First, being aware that every issue deserves detailed examination: we are surprised that examining minutiae led to such improvements and had such interesting consequences; equally, we are worried that it is too easy just to build without designing. Being aware that one can design and make explicit tradeoffs is half the battle.

Secondly, that social factors are important. In the case of **ded**, there were significant advantages in Richard's proprietary ownership of the design. 'Egoless programming' was a popular slogan of the 1970s (Weinberg, 1971): our activity in designing **ded** was quite the opposite. We don't think that

---

change it or to edit the whole line into another command. This is not equal opportunity and a bug we frequently regret.

it would have been so principled a design if Richard hadn't had absolute control over what happened to it.

Thirdly, that design evolution works: **ded** was still evolving up to about three years ago. We believe that in an innovative activity, which is what the design of **ded** was in its time, sit-down-design wouldn't have worked at all. Several times we designed an improvement, tried it out and found after some weeks of experimentation that it didn't work well in practice, so we took it out. Given that we were breaking new ground and that there were so many variables necessarily being altered at the same time, we couldn't have got anywhere without experiment. An ergonomist might have suggested some novel improvements, but we can't imagine how they could have anticipated any of our experimental 'findings'. Perhaps our view is blinkered, but it seems that first of all you need a reasonable system to start experimenting with, and that is where all the effort goes to.

Some of **ded**'s design decisions were quite arbitrary, while some were due to practical issues of implementation. For example, the difficulty of implementing a powerful editor on a small machine precluded a very complex design. Some decisions were initially arbitrary but later constrained other decisions; some, however, are independent and universally arbitrary. We think it is fair to claim that **ded** is easy to use *because it has a principled design*. Even if the principles are ergonomically awry, the user at least knows what to expect—or at least he will see the precise effect in a few milliseconds—in principle he can see *everything* pertinent to his use of the editor. No such claims have been made for any other editors, but in the final analysis it would take theories and experiments to establish claims of ease-of-use. Even so, we expect effects would be eclipsed by previous experience. If that is the case, then the sooner principled interactive system design catches on, the sooner users will cease to obtain their first experience on unprincipled systems.

John Long has proposed that the problem facing cognitive ergonomics can be characterised as how to increase compatibility between the computer's representation and the user's representations. With a declarative user interface design, the user need have no model or understanding of the history of a session with **ded**. There can be no mode changes in the past that can possibly affect the present. Editing a picture immediately increases user/computer compatibility.

Computer scientists get bored with a subject when the technology moves past it. We really don't care what happens to **ded** any more. A good word processor on a home micro is in many ways more useful than **ded** already—but not necessarily easier to use!—so hardly any users care about it either.

Those that are forced to use it (we used it to write this chapter) know that it is elderly and put up with its little quirks. We would guess that now, when ergonomists might be able to say what a text editor should *really* be like, nobody would want to listen because users want the latest technology and computer scientists want to think about the next step forward.

Appendix 1. Ded command summary

Apart from the single-keystroke direct manipulation commands (e.g., Figure 8.1), **ded** provides the following commands:

— search, search-and-replace, search-and-replace-everywhere, interrupt current search or replacement
— mark lines with a letter, copy marked region, delete marked region, move marked region, do any command within a marked region (e.g., search-and-replace), go to start or end of marked region.
— write file, read file, append file, write current file and edit another
— quit (abandon changes, or update file), pause (to resume at a later date), ok-exit
— undo (undo any number of keystrokes—including search/replace interrupts, back to the last file write)
— scroll screen by a page (up or down), go to a named or numbered line, go to first or last line
— execute UNIX commands, execute commands (e.g., compilers) and save diagnostics, go to next error line (in any file)
— various minor commands, such as: silence bell, change word-wrap behaviour, set tabs, find name of current file, redraw screen, find last usage of a given command *etc.*

Appendix 2. Why edit a picture?

In the mid 1970s we were just emerging from the Teletype™ era when most computer-readable and computer printed text was in upper case with very few different punctuation marks.[10] But UNIX had been designed to utilise the full ASCII character set—upper and lower case letters and various punctuation marks. Various conventions had been developed so that UNIX programmers could type commands to UNIX even when their terminals were deficient. On our VDTs there were no keys to print any of the characters left curly bracket '{', right curly bracket '}', vertical bar '|', tilde '~' nor opening quote ' '. The terminal didn't recognise commands to print those characters either. The first three at least of the characters were essential in order to program in **C**, the main UNIX programming lan-

---

[10]UNIX practice is still to refer to terminal connections as *ttys*, with obvious etymology.

guage, so there was a real problem when constructing, editing, printing or otherwise reviewing program texts.

The convention we used at QMC was to type a left curly bracket as two characters '\('—backslash followed by left bracket. In just the same way a text would be printed with '\(' in every position where '{' should have appeared. This seemed quite rational, but for reasons of 'consistency' since the two marks in the printed text represented *one* character, the two keystrokes were understood to have constructed only *one* character. Under the normal convention that RUBOUT deletes the last character typed, the sequence of keystrokes a, b, \, (, RUBOUT, c would be understood as 'abc' since the RUBOUT deleted the curly bracket you had typed. All very well, so far as it goes.

The designers of this convention had had some difficulty in putting it into practice. In particular, perhaps in order to reduce the number of key depressions, they had taken the totally daft decision that backslash was only part of a composite character when it needed to be. So '\a' was interpreted as a *two* character sequence—backslash followed by a—but '\(' was the *single* character '{'. The consequence was that a line which contained the four characters 'a\(b'—a, backslash, left bracket, b—would be printed in *exactly* the same way as a line containing the three characters 'a{b'. More amazing still was that backslash followed by RUBOUT meant 'insert the RUBOUT key code in the line'. It was possible to type a line which contained a backslash followed by a left bracket: you typed \, SPACE, RUBOUT, (. The RUBOUT eliminated the space and left the backslash as a single character, which by now the system has forgotten about. You have now typed a line that means \(but looks exactly like what you would have if you had simply typed '\(', which actually means {. In other words, the meaning depends on what you did, not on what it looks like. This is history sensitive, or imperative; it is not declarative, it is certainly not editing a picture—and a bad thing for users.

All this was so confusing that the decision was made that **ded** wouldn't edit an ASCII text but a *picture* of the ASCII text. In that picture a left curly bracket would appear as *two* characters and would *behave* as two characters. When you had finished working on the text the picture would be translated back into ASCII: note that the picture is translated, not how you made the picture (which would be being history sensitive).

This decision to work on a picture of the original text, not the text itself, opened up lots of possibilities. It made it possible to explain where the editing position went when you moved it up, down, left or right—it went up, down, left or right in the picture without worrying whether the cursor was exactly over an ASCII character (what would have happened if the cursor had been moved up in between '\' and '(', or—worse—moved beyond the end of the line where there were *no* ASCII characters at all?). The cursor in **ded** could be moved to any of these positions if you wanted it to be there.

It was an enormous simplification. Layout characters—line feeds, carriage returns, tabs—could all be ignored once the input had been translated into a picture. Its drawback was that the translation from ASCII text to picture was many-to-one, so information was lost. In particular tab layout information was lost, blank spaces at the end of lines disappeared and, at least in the early stages of **ded**'s evolution, so did blank lines at the end of a text. But the simplicity of the explanation overcame most people's objections. **Ded** is what is nowadays known as a 'WYSIWYG' editor, from the phrase 'what you see is what you get'. This is, though, a slightly idiosyncratic use of the phrase: for (apart from programmers) few users would ever want to see on paper what they had got, they would rather pump it through a text formatter first.

Appendix 3. Ded's search notation

Search notations are often based on so-called regular expressions. Many text editors use a period '.' (or even an invisible control character) to describe 'match any character' and an asterisk '*' to describe repetition. For example, the following **ed** command would find a place in a text where the word 'the' was followed after some number of characters by the word 'stop':

```
/ the .* stop /
```

There are several problems associated with using regular expressions for specifying searches in natural language text; we mention three, our design of **ded** solves two. First, the commands specify *exact* matches: editors often have various conventions e.g., to permit searching for upper and lower case letters conveniently. This causes the second problem, to provide worthwhile utility *lots* of special symbols are needed—and the user might use them by accident or incorrectly. Thirdly, since period and asterisk (and lots of other characters) have special meanings it is necessary to use a conventional description of them: e.g., to find a line which actually contained a period followed by an asterisk—like the line in the illustration above—it was necessary to type the command

```
/\.\*/
```

This *overloads* many characters: the special characters have *two* meanings, themselves and themselves-as-special-symbols. This can cause confusion for the user, especially since the characters are likely to be infrequently used ones such as '@', '*', '[' and '{'—so the user can easily forget that they have a special meaning. We were often tripped up by this convention because we often wanted to look for particular **nroff** commands in text, and **nroff** commands always start with a period (and they have lots of backslashes in them as well which cause similar problems we needn't go into). But in any case the novice shouldn't have to learn to do something complicated just to save the expert a few keystrokes. So we decided that in **ded**'s search patterns period an asterisk and all the rest should have no special meaning. That made one less thing to explain to novices.

In order to describe repetitions and such like we turned the standard convention on its head. A period preceded by a prime means 'match any character' and an asterisk preceded by a prime means repetition. So in **ded**'s notation the two searches illustrated above become

```
/ the '.'* stop /
/.*
```

The advantage of this approach is threefold: the probability of accidental use of a special symbol is minimised; typing errors resulting in invalid symbol combinations are easily detected (e.g., 'x has no meaning); many special symbols can be provided. Thus **ded** provides a simple searching language for naïve users (with little chance of their invoking advanced features accidentally), and at the same time it is able to provide many search and replacement features found on no other editors.

Anyway, the notation works, and it works well enough.

Bibliography

J. M. Carroll (1987), *Paradox and the Active User: Implications of Learning for Design*, Complex Learning Workshop, University of Lancaster.

R. E. Granada, R. C. Teitelbaum & G. L. Dunlap (1982), *Effects of VDT Command Line Location on Data Entry Behaviour*, Proceedings Human Factors Society.

W. J. Hansen (1971), *User Engineering Principles for Interactive Systems*, AFIPS Conference Proceedings, **39**, pp523–532, Las Vegas.

N. Meyrowitz & A. van Dam 1982, *Interactive Editing Systems*, ACM Computing Surveys, **14**(3), pp321–415.

D. L. Parnas, P. C. Clements (1986), *A Rational Design Process: How and Why to Fake It*, IEEE Transactions on Software Engineering, **SE-12**(2), pp251–257.

R. Pike (1987), *The text editor sam*, Software—Practice and Experience, **17**(11), pp813–845.

C. Runciman & H. W. Thimbleby (1986), *Equal Opportunity Interactive Systems*, International Journal of Man-Machine Studies, **25**(4), pp439–451.

R. M. Stallman (1984), *EMACS: The Extensible, Customizable, Self-Documenting Display Editor*, in Interactive Programming Environments, eds. D. R. Barstow, H. E. Shrobe & E. Sandewall, pp300–325, McGraw-Hill Pub.

B. Sufrin (1981), *Formal Specification of a Display Editor*, PRG–21, Programming Research Group, Oxford University.

H. W. Thimbleby (1981), *A Word Boundary Algorithm for Text Processing*, Computer Journal, vol. **24**, pp249–255.

H. W. Thimbleby (1982), *A Text Editing Interface: Definition and Use*, Computer Languages, vol. **7**, pp25–40.

H. W. Thimbleby (1983), *Guidelines for 'Manipulative' Editing*, Behaviour and Information Technology, vol. **2**, pp127–161.

H. W. Thimbleby (1984), *Generative User-Engineering Principles for User Interface Design*, Proceedings First IFIP Conference on Human Computer Interaction, INTERACT'84, London, volume **2**, pp102–107.

G. M. Weinberg (1971), *Psychology of Programming*, Van Nostrand Reinhold, London.

Acknowledgements

The authors gratefully acknowledge helpful criticism from Tony Fisher, David Laukee, John Long, and Andy Whitefield.

Index